10分鐘OK燜燒料理

研出版

作者序
Preface

看著身邊的朋友由拍拖、結婚，然後有了小朋友，組成一個個快樂的小家庭；
由以前從不入廚，到現在常常將家人的飲食健康掛在嘴邊。看著朋友的改變，
令我這幾年間鼓起了一份使命感，常常想為他們寫下更多的健康食譜給他們
參考。朋友們的要求，除了想菜式健康一點，也當然想要一些烹調方法簡單
和快捷的食譜。我相當明白，因為於現今生活中，大家都很忙碌，要將在家
入廚的準備事情簡化是必然的大趨勢。

我也知道這些需求不只屬於我的朋友，相信拿起這本書的你都有相同的要求。

早幾年前，我接觸了「燜燒罐」。跟大部份的讀者一樣，第一次接觸燜燒罐
煮食時，腦裡都有不少的問號，始終不相信一隻杯居然可以煮出不同菜式。
但當我用了十數次以後，便開始相信它，甚至愛上用這個小罐來協助煮食。
在我手中，燜燒罐的功用不會只是用來保溫或烹煮小朋友的副食品，這些細
小可愛的罐子其實既能煮特色粥飯，亦可煮出家常小菜。我將這幾年的烹調
心得和經驗整合成書中的貼士和食譜，和大家一一分享。

至於這本書 Chapter 5-7 內推介的「真空煲」，我真的怪自己太遲才認識它
們！還記得小時候我和親戚們，每個家庭總有一個真空煲，但它們全都是銀
色的、外型體積都很大，我的婆婆最喜歡用它們來煲湯、燜餸，烹調份量龐
大的菜式。直到一年前，Thermos 的真空煲加入了很多不同顏色，外型變
得時尚簡潔，部分體積纖形的款式，特別適合一家四口的家庭使用，令人對
它們的感覺煥然一新。除了外型之外，新款的真空煲內涵依舊吸引，方便易
用程度實在一流。只需要利用 5-10 分鐘在爐上烹煮，然後放在一旁待約 30
分鐘的燜焗時間，出來的效果等同我們平時要在爐上煮 2-3 小時的菜式一
樣。這麼省能源、方便又少煙火的烹調方法；你說怎能不會愛上它？

這一本書內的菜式食譜以香港家常菜為根底，希望你們都能在家輕鬆煮出一
道道的窩心家常菜。

Hilda Leung

序
Preface

對現代人來說，真空保溫產品似乎並不陌生。從隨身攜帶的保溫瓶、燜燒罐，家用的真空煲以至到手斟壺，仔細一看其實四周都充滿著它們的身影。縱然它們型態各異，功能不盡相同，但我們身邊總少不了這種好幫手。

然而，很多人卻看輕了這個小幫手的可塑性，認為它們就只能為我們提供暖暖的熱水，令它們未能完全發揮潛能！一直以來，香港膳魔師都致力鼓勵活用燜燒罐及真空煲，一邊提倡節省能源，保護環境，一邊亦能為市民帶來更便利的生活。眼看愈來愈多的市民，於不同的平台，交流他們對燜燒罐及真空煲的使用心得，他們創意與想像，以至對我們產品的熱愛，都令我們大為感動。所以我們今次專程找來膳魔師香港區的代言人梁雅琳小姐，為我們製作一本專為香港人口味而設的食譜，將各種使用燜燒罐的方法及技巧輯錄成書，與您們分享。希望能讓香港人在繁忙的生活中，騰出更多珍貴的時間，享受生活。

希望現在手執這書的您，能從書中體會到使用燜燒罐的樂趣及方便，亦同時在使用燜燒罐的過程中，不知不覺地就能夠為地球節省能源，為環保出一分力。

膳魔師 (香港) 有限公司

目錄 Contents

Chapter 4
香焗米飯

Chapter 5
窩心湯品

Chapter 6
滋味燜餸

Chapter 7
新穎粥品糖水

認識燜燒罐的原理與構造

初次使用燜燒罐？其實它是透過優良的真空保溫效能，讓食物能夠在短時間內燜煮妥當。它的構造是怎樣？讓我們來看一看吧！

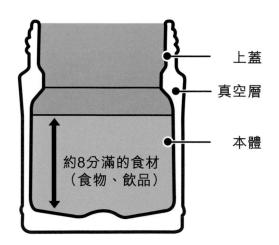

上蓋

真空層

本體

約8分滿的食材
（食物、飲品）

燜燒罐在外層與內層之間，設有真空層，能有效阻止熱力的流失，這樣只要放入足夠高溫的食物，就能確保長時間的高溫來讓食物熟透，也能讓食物保持熱度，攜帶外出也能享用熱騰騰的美食了。

真空保溫原理

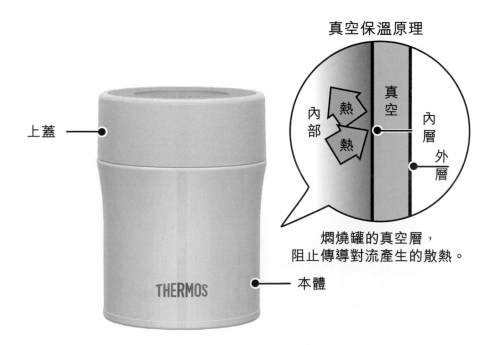

上蓋

內部

熱

熱

真空

內層

外層

本體

THERMOS

燜燒罐的真空層，
阻止傳導對流產生的散熱。

Tips

燜燒罐正確使用步驟

想安全又放心地使用燜燒罐？秘訣在於正確地做好每一道食物處理步驟，讓我們快速看一遍吧！

1
預熱
注入沸水（100℃）至罐內三分滿，加蓋搖晃，先將燜燒罐內部的溫度提升。

4
瀝水
打開罐蓋，瀝去所有氽燙食材的水。

2
放入食材
放入需要烹煮時間最長的食材，例如肉類。

5
加入餘下食材
加入食譜中其餘較易煮熟的食材，例如蔬菜。

3
氽水
注入沸水並扭上蓋子搖晃，氽燙食材。

6
靜置燜焗
注入沸水後，加蓋後略搖晃，利用燜燒罐良好的保溫功能，將食材焗「煮」至熟。

Tips

不失敗的燜燒秘訣

使用燜燒罐來做料理時，偶然也會出現失敗的情況？明明依足了步驟，那就可能是沒有好好注意當中的細節，這裡給您一點窩心的小貼士吧。

1. 裝載比例
留意，最多只可以於罐內注入沸水至八分滿，這樣才有足夠的空間於加蓋後形成燜燒狀態。

2. 食材大小
使用燜燒罐煮食的食材大小也要留神：蔬菜類食材切粒，約 2 厘米大小；肉類食材切粒，約 2 厘米大小或切成薄片。

3. 切勿中途開蓋
在靜置燜燒中途，燜燒罐切記不能打開，此舉會令罐內溫度下降，影響燜燒食材的效果，令食材不夠熟透。

4. 改用熱湯
使用燜燒罐靜置燜焗前注入的滾水，也可以用熱湯代替，這樣便令菜式味道更豐富及美味。

5. 沸水溫度
注入燜燒罐內的沸水或湯，必需達至 100℃，這點也往往容易被忽略哦！

6. 汆水的重要性
經過汆水能確保所有食材有足夠的時間和空間受熱，於燜燒時能充分「煮」熟。

注意事項

1. 請勿將燜燒罐放入焗爐、微波爐、燒烤爐等電器用品中使用。

2. 請勿將燜燒罐放近於高溫地方，免導致變形、變色、印漆脫落。

3. 請勿將燜燒罐的配件放入沸水中煮沸，以免導致配件變形、滲漏、污染。

4. 避免掉落、碰撞或強烈撞擊，以免導致變形、損毀而影響其保溫功能。（下頁續）

Tips 不同食材的注意事項

燜燒罐用來製作各類食譜也很方便，但若食譜裡需要使用某些食材時，就得多加留意處理的程序了！

1. 乳製品
避免放入乳製品、牛奶等，放置時間過長可能會導致腐壞。

2. 酸性食材
避免放入檸檬汁、酸梅汁等酸性飲品及容易腐壞的生食食物，以免影響燜燒罐保溫功能而導致食物變質。

3. 改變內壓的食材
避免放入乾冰、碳酸飲料，以免罐內內壓上升，導致不能開啟上蓋、噴出罐內食物或上蓋損毀等危險問題。

4. 鹽分較多的食材
若放入有鹽分的食物或湯水，建議於 6 小時內食用。燜燒罐不可當作貯存食物盒使用，要先將剩餘的食物倒出，放入雪櫃冷藏。

5. 定期檢查
定期檢查燜燒罐內的矽膠部件，如出現損耗現象請聯絡香港膳魔師查詢。

注意事項

5. 請勿使用稀釋劑、揮發油、金屬刷或研磨粉精洗燜燒罐，以免導致磨損、生銹等現象。

6. 請勿使用漂白劑清洗燜燒罐的裡、外兩面，以免影響其保溫、保冷功能或導致印漆脱落。

7. 清洗時，請勿將燜燒罐浸泡在水中，若水份滲入了金屬、塑料之間的接合位，可能導致生銹，影響其保溫、保冷功能。

8. 請勿將熱食貯存在燜燒罐內過久，盡量於 6 小時內食用，避免食物過時變質腐壞。

真空煲正確使用步驟

真空煲也是非常實用與靈活性高的煮食器材，只要煮沸食材後再進行燜焗，就能快速減省煮食時間，對於繁忙的上班族來說實在太方便了！

真空煲的煮食步驟

1. 熱力烹調

取出真空煲內鍋，放於電磁爐或明火上，加入食材以中至小火烹調至沸騰(100℃)。

- 湯汁至少要蓋過食材。
- 建議食材放至內鍋八分滿（能達至最佳煮食效果）。

2. 加蓋煲煮

當鍋內食材煮滾後，加蓋繼續烹調。

- 煲煮時間：需要因應不同菜式而作調整。
- 減少注入水份：相較平時的燜煮方法，真空煮食較能保存菜式水份，保持食材的原汁原味。

3. 保暖燜燒

將真空煲內鍋放入外鍋內燜燒，以真空煲的良好保溫功能繼續「烹調」。

- 此時若打開真空煲的蓋，會令熱力流失，令真空煲不能達至最佳煮食效果，需要再次取出真空煲內鍋加熱至沸騰後，放回真空煲外鍋內。

4. 菜式完成

完成的菜式可以放於真空煲外鍋保溫。

- 若真空煲內裡食材份量太少，能保持一定溫度的時間會較短，所以進食前要徹底再次加熱。
- 保溫時間過長，食物有機會出現變壞的情況，所以不建議用真空煲作保存食物之用。

清潔保養方法

1. 使用完畢請立即清洗乾淨並保持乾燥。

2. 建議先用溫水稀釋食器用中性清潔劑，並使用清潔海綿將整個內鍋清洗乾淨，拭乾水份後保持乾燥。切勿使用氯系漂白劑。

3. 如果長時間不使用，請將鍋具清洗乾淨，充分乾燥後，收藏在陰涼乾燥處。

4. 請用沾有稀釋食器用中性清潔劑的布清潔燜燒外鍋及燜燒外鍋上蓋，再用清潔的布擦拭，最後用乾布將水份完全擦乾。切勿使用漂白劑。

5. 可將固定於上蓋的矽膠用水直接沖洗或浸泡後沖洗，切勿放於沸水之中加熱。使用後請立即從燜燒鍋外鍋上蓋除下，清洗及乾燥後再將矽膠裝回。

注意事項

1. 請勿將真空煲放入焗爐、微波爐、燒烤爐等電器中使用。

2. 請勿直接將食材放入真空煲外鍋，以免造成生銹、腐蝕。

3. 請勿將重物放在真空煲上，以免造成變形、破損。

4. 請勿將真空煲外鍋加熱，由於外鍋底部為塑膠物料，加熱或會引致火災。

5. 如曾將內鍋連食材冷藏，請確保食材待至室溫後才好加熱，以免食物在加熱時溢出。

常見產品問題

問題：無法燜燒及保溫效果不好

原因：食物量太少

解決方法： 如果食物量太少，鍋內溫度容易下降，使燜燒效果不佳。請增加食物分量至八成滿或再次加熱內鍋。

原因：反覆開啟燜燒中的真空煲外鍋或內鍋上蓋

解決方法：多次開啟燜燒中的真空煲外鍋或內鍋上蓋，會使鍋內溫度下降。請再次加熱內鍋。

原因：長時間燜燒或保溫

解決方法： 食用前請再次加熱內鍋至沸騰 (100℃)。

問題：發出異味

原因：污垢附著

解決方法： 每次使用後請立即清洗乾淨，避免食物殘留於煲內，並保持乾燥。

問題：真空煲內鍋變色

原因：真空煲內鍋燒焦

解決方法： 請將內鍋放於加有食器用中性清潔劑的溫水中稍微浸泡，再用海綿搓洗。

問題：外鍋內側有水滴附著

原因：真空煲內鍋燒焦

解決方法：真空煲外鍋與內鍋內側的溫差可能會引致結露，屬於正常現象，使用後請將水滴擦乾。

燜燒罐與真空煲 Q&A

燜燒罐

1. 我的燜燒罐尺寸跟食譜寫的不一樣可以用嘛？

本書會註明適合的尺寸，你可以按照比例增減，不過建議以食譜上的尺寸烹調為佳，若使用不同尺寸的燜燒罐或擔心食材難以煮熟，可將食材略煮至沸騰後，再放進燜燒罐燜煮。

2. 怎樣才是適合的食材大小呢？

輕、薄、短、小是大原則，食物盡量切小 (以 2 公分為佳)、剪細時為 3-5 公分，麵條則控制在 7 公分左右，這樣才能方便導熱。

3. 燜不熟怎麼辦呢？

相信這是大家都很擔心的問題，切記，不要因為擔心而中途就打開來查看，因為燜燒罐是運用保溫導熱來進行烹煮，打開後熱氣會散出，溫度就會下降，而且燜燒罐開罐後散熱速度很快，亦會因為食材狀況跟食物分量及大小而改變，不過熟悉之後，就會很清楚怎麼使用。也可以在「預熱」或「燜煮」時，上下搖動燜燒罐，使熱度均勻。

4. 「預熱」作用是甚麼呢？還有預熱後的食物要一起倒出來嗎？

「預熱」是為了讓燜燒罐隨時保持在 100℃ 烹煮食物，也因此料理步驟比較複雜時，通常會希望食物先行「預熱」，步驟多數依照食物的易熟度來遞減分配。預熱後食物留在罐內，再增加其他食材一起預熱即可。

5. 為甚麼燜燒罐有時候會打不開？

打不開往往是因為內外罐溫差的關係，通常發生在早期的燜燒罐沒有內蓋設計，如有此問題，在外蓋沖冷水，稍微擦拭後打開即可，或可致電香港膳魔師客戶服務熱線查詢。

真空煲

1. 本書使用的真空煲尺寸是多少呢？

本書使用 3 公升 - 8 公升的真空煲，如使用不同尺寸的真空煲，食材請作比例增減即可。要注意如將真空煲用作燜燒，食材應以超過內鍋一半及不足八分滿為最理想的烹煮方式。

2. 燜不熟該怎麼辦？

切記，不要因為擔心而中途就打開真空煲查看，因為真空煲是運用高保溫效力來進行烹煮，打開之後會流失熱氣，使鍋內溫度下降。雖然真空煲的退溫速度不比燜燒罐快，而本書食譜當中，也有幾次需要開蓋加溫的情況，但那都是在料理將近完成時作最後加工。如有不放心處，在開鍋查看後，建議將內鍋再次放回火源加熱至沸騰 (100℃) 再進行燜煮。

保固聲明

根據保固內容膳魔師（香港）有限公司僅保證本公司香港，澳門地區出售之膳魔師系列市商品，產品保養事宜必須提供有效之購買單據証明。

自購買日起正常使用情況下各類產品保固期限如下：
- 真空保溫瓶之真空本體的保固期為 1 年。
- 3 公升或以上的真空燜燒鍋之真空本體的保固期為 5 年。
  * RPF-20 不銹鋼真空煲保固期為 1 年。

保固範圍

- 如屬人為過失或天災地變等不可抗拒之外力所引起的損壞，則不在保固範圍內。
- 零配件及其他小五金商品屬消耗品不在保固之內，請定期檢查。
- 基於衛生理由吸管及配件不設退換服務。
- 塑膠及矽膠等配件為消耗品，正常使用下大約一年即須檢查，如有表面損傷或不平整的狀況，屬於正常損耗現象， 請購買新配件更換。

如有任何保養事宜，請直接致電或電郵致香港膳魔師公司：
電話：2608 0880
傳真：2608 0886
e-mail: customer-service@thermos.com.hk

Food Jar

Chapter 1

滋味甜品

不用開爐,只需熱水便可製作出滋潤甜品,方便易做的食譜,連小朋友都可以輕易學懂!

雪梨紅棗馬蹄糖水

雪梨香甜清熱，加入馬蹄粉為此食譜糖水增添口感。

JCU-500-P

 材料

雪梨（去皮，去籽，切粒） · · 1/2 個

紅棗（用水浸軟，去核，切件） · · 2 粒

馬蹄（去皮，切粒） · · · · · 2 粒

馬蹄粉 · · · · · · · · · 1 湯匙

糖 · · · · · · · · · · 1.5 湯匙

 做法

1 燜燒罐內注入熱水預熱，加蓋，略搖晃，靜置 1 分鐘，倒水。

2 加入雪梨、紅棗和馬蹄，注入熱水蓋過食材表面，拌勻後加蓋，略搖晃，靜置燜焗 15 分鐘，瀝去水分。

3 下馬蹄粉和糖，邊攪拌邊注入熱水至罐內八分滿，加蓋，略搖晃，靜置燜焗 5 分鐘，完成。

Tips

除了可使用雪梨來製作，亦可選用同樣香甜爽口的蘋果代替。

薑汁蜜紅豆芋頭

選用簡單方便的日式罐頭蜜紅豆，為薑汁芋頭糖水加添香甜味，適合冬天滋補、暖胃的糖水便完成了。

JBJ-302 (P-P)

 材料

芋頭（去皮，切粒）· · · · ·	30 克
日式蜜紅豆· · · · · · · ·	適量
薑汁 · · · · · · · · ·	1 湯匙

 做法

1 燜燒罐內注入熱水預熱，加蓋，略搖晃，靜置 1 分鐘，倒水。

2 下芋頭，注入熱水蓋過表面，拌勻後加蓋，略搖晃，靜置燜焗 5 分鐘，瀝去水分。

3 加入日式蜜紅豆和薑汁，注入熱水至罐內八分滿，拌勻後加蓋，略搖晃，靜置 5 分鐘，完成。

草莓生芋絲凍

日式生芋絲不一定用作鹹食,將生芋絲放入雪櫃冷藏一下變成凍芋絲,然後配上香甜士多啤梨醬可有另一番風味。

JBJ-302(P-Y)

 材料

士多啤梨（去蒂,沖淨,切粒）　　80 克

蜜糖 · · · · · · · · · · · 1 湯匙

冰塊 · · · · · · · · · · · 1 粒

日式生芋絲 · · · · · · · 120 克

做法

1 取 60 克士多啤梨、蜜糖和冰塊放入攪拌機內攪拌,盛起備用。

2 燜燒罐內注入熱水預熱,加蓋,略搖晃,靜置 1 分鐘,倒水。

3 下日式生芋絲,注入熱水至罐內八分滿,拌勻後加蓋,略搖晃,靜置燜焗 3 分鐘,瀝去水分,倒出蒟蒻條,放涼。

4 注入冰水冷卻燜燒罐,加蓋,略搖晃,靜置 1 分鐘,倒去冰水。

5 加入已放涼的日式生芋絲、士多啤梨醬和餘下士多啤梨,拌勻後加蓋保冷,即可隨時享用。

桃膠粟米木瓜糖水

粟米汁和木瓜的天然甜味,足以使菜式不加糖,
也有香甜的好味道。

JBM-500-MNT

 材料

桃膠 (浸水一晚,洗淨,瀝水) · ·	5 克
粟米粒 · · · · · · · · · · ·	20 克
木瓜 (去皮,切粒) · · · · ·	25 克

 做法

1 將 15 克粟米粒和 100 毫升水放入攪拌機內攪拌,隔渣後取 50 毫升粟米汁備用。

2 燜燒罐內注入熱水預熱,加蓋,略搖晃,靜置 1 分鐘,倒水。

3 放入桃膠、粟米汁、木瓜和餘下粟米粒,注入熱水至罐內八分滿,拌勻後加蓋,略搖晃,靜置燜焗 25 分鐘,完成。

 Tips

桃膠需要先用水浸發過夜,然後可以倒入篩中攪拌,並挑走黑色雜質。

椰汁紫薯小米糖水

小米是五穀雜糧類中，唯一屬鹼性食物，容易被消化及烹煮，用作焗飯或糖水都十分適合。

JBJ-302 (P-B)

 材料

紫薯（去皮，切粒） · · · · ·	35 克
小米 · · · · · · · · ·	1 湯匙
椰汁 · · · · · · · · ·	1 湯匙
蜜糖 · · · · · · · · ·	1/2 湯匙

做法

1 燜燒罐內注入熱水預熱，加蓋，略搖晃，靜置 1 分鐘，倒水。

2 加入紫薯，注入熱水蓋過表面，拌勻後加蓋，略搖晃，靜置 5 分鐘，瀝去水分。

3 加入小米，注入熱水至罐內八分滿，拌勻後加蓋，略搖晃，靜置燜焗 5 分鐘，開蓋，加入椰汁和蜜糖拌勻，完成。

Food Jar

Chapter 2
惹味家常菜

善用保溫力極佳的燜燒罐,輕易「煮」出
一道道充滿住家味道的菜式!即使身處辦
公室,都可輕易變出不同的款式。

蠔油雲耳豚肉翠玉瓜

簡單的三種食材,已經能「煮」出惹味拌飯的菜式。

JBJ-300-MNT

 材料

豚肉片(切細) · · · · · · 50 克

翠玉瓜(洗淨,切薄片) · · · 40 克

雲耳(浸軟,去蒂,切細) · · · · 1 克

 調味料

鹽 · · · · · · · · · · · 適量

蠔油 · · · · · · · · 1 湯匙

 做法

1 燜燒罐內注入熱水預熱,加蓋,略搖晃,靜置 1 分鐘,倒水。

2 放入豚肉片,注入熱水蓋過表面,拌勻後加蓋,略搖晃,靜置 2 分鐘,瀝去水分。

3 再加入翠玉瓜和雲耳,注入熱水至罐內八分滿,拌勻後加蓋,略搖晃,靜置燜焗 5 分鐘,瀝去水分。

4 最後,加入調味料拌勻,完成。

Tips

使用筷子將豚肉片充分散開,確保每一片都能均勻受熱。

麵豉苦瓜燜肉丁

經過燜焗的薄片苦瓜，已去除大部分的甘苦味，
不喜愛苦瓜的人仕亦可一試。

SK3020WH

 材料

豬肉（洗淨，切粒） · · · · ·	50 克
苦瓜（洗淨，切薄片） · · ·	70 克
鮮冬菇（沖淨，去蒂，切片） · · ·	2 隻
日式醃漬牛蒡 · · · · · · ·	適量

調味料

鹽 · · · · · · · · · · ·	適量
糖 · · · · · · · · · · ·	適量
麵豉醬 · · · · · · ·	1/2 湯匙
水 · · · · · · · · ·	1/2 湯匙

做法

1 燜燒罐內注入熱水預熱，加蓋，略搖晃，靜置 1 分鐘，倒水。

2 放入豬肉，注入熱水蓋過表面，拌勻後加蓋，略搖晃，靜置 5 分鐘，瀝
去水分。

3 再加入苦瓜和鮮冬菇，注入熱水至罐內八分滿，拌勻後加蓋，略搖晃，
靜置燜焗 5 分鐘，瀝去水分。

4 最後，加入調味料和日式醃漬牛蒡拌勻，完成。

黑松露法邊豆蝦仁

加入少許黑松露醬，即可令平凡的鮮蔬蝦仁
充滿香氣。

JBM-500-BK

 材料

蝦仁 · · · · · · · · ·	12 隻
法邊豆（洗淨，切段） · · · ·	70 克
黃甜椒（洗淨，去蒂，切條）· ·	1/3 個

調味料

鹽 · · · · · · · · · ·	適量
黑松露醬 · · · · · · ·	1/2 茶匙
橄欖油 · · · · · · · · ·	適量

 做法

1 燜燒罐內注入熱水預熱，加蓋，略搖晃，靜置 1 分鐘，倒水。

2 放入蝦仁，注入熱水蓋過表面，加蓋，略搖晃，靜置 1 分鐘，瀝去水分。

3 再加入法邊豆和黃甜椒，注入熱水至罐內八分滿，略搖晃，靜置燜焗 3 分鐘，瀝去水分。

4 最後，加入調味料拌勻，完成。

☕ Tips

因為蝦仁含有豐富蛋白質，當遇上熱水後可能會浮出白泡，所以可以選擇
再加入熱水「汆水」以去除泡沫。

XO 醬魚片茄子

XO 醬為魚片菜式加添辣味，令人更開胃。

JBM-500-GRP

 材料

龍脷柳（斜切薄片）	· · · · ·	80 克
茄子（洗淨，切條）	· · · · ·	1/2 條
荷蘭豆 （洗淨，去邊）	· · · ·	20 克

調味料

鹽	· · · · · · · · ·	適量
糖	· · · · · · · · ·	適量
XO 醬	· · · · · · ·	1/2 茶匙
蠔油	· · · · · · · ·	1/2 茶匙

做法

1 燜燒罐內注入熱水預熱，加蓋，略搖晃，靜置 1 分鐘，倒水。

2 放入龍脷柳，注入熱水蓋過表面，拌勻後加蓋，略搖晃，靜置 3 分鐘，瀝去水分。

3 再加入茄子和荷蘭豆，注入熱水至罐內八分滿，拌勻後加蓋，略搖晃，靜置燜焗 5 分鐘，瀝去水分。

3 最後，加入調味料拌勻，完成。

橙蜜泡菜蜜豆帶子

橙醬的甜味中和了泡菜的辛辣，令菜式甜中帶辣，更添吸引。

JBJ-302(P-B)

 材料

帶子	30 克
蜜豆（洗淨，去邊）. . . .	30 克
粟米芯（沖淨，切細）. . .	30 克
泡菜（切細）.	30 克

 調味料

橙醬	1 茶匙
鹽	適量

 做法

1 燜燒罐內注入熱水預熱，加蓋，略搖晃，靜置 1 分鐘，倒水。

2 放入帶子，注入熱水蓋過表面，拌勻後加蓋，略搖晃，靜置 3 分鐘，瀝去水分。

3 再加入蜜豆和粟米芯，注入熱水至罐內八分滿，拌勻後加蓋，略搖晃，靜置燜焗 3 分鐘，瀝去水分。

4 最後，加入泡菜和調味料拌勻，完成。

☕ Tips

額外加入香甜的鮮橙粒，可為菜式帶來更豐富的口感。

冬菇豆卜炆豚肉粒

拌入蒜蓉豆豉醬令菜式更俱香港風味，
增加住家菜的感覺。

JBM-500-CRB

 材料

豬肉（切幼粒）· · · · · · ·	65 克
豆卜（沖淨，切半）· · · ·	20 克
鮮冬菇（沖淨，去蒂，切半）· · ·	2 隻
秀珍菇（沖淨，撕成小塊）· ·	20 克
娃娃菜（洗淨，切細）· · · ·	2 片

調味料

糖 · · · · · · · · · · ·	適量
蒜蓉豆豉醬 · · · · · · ·	1 茶匙

做法

1 燜燒罐內注入熱水預熱，加蓋，略搖晃，靜置 1 分鐘，倒水。

2 放入豬肉，注入熱水蓋過表面，拌勻後加蓋，略搖晃，靜置 5 分鐘，瀝去水分。

3 再加入豆卜、鮮冬菇、秀珍菇和娃娃菜，注入熱水至罐內八分滿，拌勻後加蓋，略搖晃，靜置燜焗 5 分鐘，瀝去水分。

4 最後，加入調味料拌勻，完成。

泰式咖喱羅勒雞丁

新鮮的羅勒令咖哩帶有清新的香氣，
菜式味濃但不膩。

JBM-500-MNT

 材料

雞肉（沖淨，切粒）· · · · ·	80 克
椰菜花（洗淨，切細）· · · ·	20 克
西蘭花（洗淨，切細）· · · ·	20 克
甘筍（洗淨，去皮，切薄片）· ·	20 克
羅勒（取葉，切細）· · · · · ·	適量

調味料

青咖喱醬 · · · · · · · ·	1/2 湯匙
椰奶 · · · · · · · · ·	1 茶匙
鹽 · · · · · · · · · ·	適量

做法

1 燜燒罐內注入熱水預熱，加蓋，略搖晃，靜置 1 分鐘，倒水。

2 放入雞肉，注入熱水蓋過表面，拌勻後加蓋，略搖晃，靜置 3 分鐘，瀝去水分。

3 再加入椰菜花、西蘭花和甘筍，注入熱水至罐內八分滿，拌勻後加蓋，略搖晃，靜置 5 分鐘，瀝去水分。

4 最後，加入羅勒和調味料拌勻，完成。

茄香迷你五香蝦丸

外形迷你，充滿天然香料味的蝦丸，配上開胃的
酸甜茄醬，相信能輕易滿足家中的大小肚皮。

JBJ-302 (P-P)

 材料

番茄（洗淨，去蒂，切細）	1/2 個
泰國蘆筍（沖淨，切段）	20 克
洋蔥（去皮，切粒）	20 克
茄膏	1/2 湯匙
鹽	適量
糖	適量

蝦丸材料 （約 5 粒）

蝦仁（切細）	50 克
白胡椒粉	適量
鹽	適量
五香粉	少許
粟米粒	5 克

 做法

1 準備蝦丸：蝦仁、白胡椒粉、鹽和五香粉放入攪拌機內攪拌，盛起後加
入粟米粒拌勻，搓成小丸子狀。

2 燜燒罐內注入熱水預熱，加蓋，略搖晃，靜置 1 分鐘，倒水。

3 注入熱水至罐內八分滿，放入蝦丸，拌勻後加蓋，略搖晃，靜置 5 分鐘，
瀝去水分。

4 再加入番茄、泰國蘆筍和洋蔥，注入熱水至罐內八分滿，拌勻後加蓋，
略搖晃，靜置燜焗 3 分鐘，瀝去水分。

5 最後加入茄膏、鹽和糖拌勻，完成。

Tips

注入熱水後才輕輕放入蝦丸，便能保持蝦丸的丸子形狀。

Food Jar

Chapter 3
開胃麵食

賣相吸引又惹味的簡易麵食,大小朋友都
會喜歡,不論野餐、還是充當午餐都毫無
問題。

欖菜肉碎烏冬

自家製作簡便的潮洲風味菜,加入橄欖菜便輕易令肉碎烏冬變得更惹味。

SK3020BK

 材料

豬肉碎 · · · · · · · · ·	50 克
粟米粒 · · · · · · · · ·	20 克
豆角(洗淨,切細) · · · · ·	1 條
烏冬 · · · · · · · · ·	1/2 個

調味料

橄欖菜 · · · · · · · · ·	1 湯匙
鹽 · · · · · · · · ·	少許
糖 · · · · · · · · ·	少許

做法

1 燜燒罐內注入熱水預熱,加蓋,略搖晃,靜置 1 分鐘,倒水。

2 放入豬肉碎,注入熱水蓋過表面,拌勻後加蓋,略搖晃,靜置 3 分鐘,瀝去水分。

3 再加入粟米粒、豆角和烏冬,注入熱水至罐內八分滿,拌勻後加蓋,略搖晃,靜置燜焗 5 分鐘,瀝去水分。

4 最後,加入調味料拌勻,完成。

香蒜鮮蝦撈麵

加入多款不同蔬菜後，簡易的撈麵便具備了不同的口感和營養。

JBJ-300-VAN

 材料

蝦仁（洗淨）・・・・・・・	30 克
椰菜（洗淨，切條）・・・・	20 克
木耳（浸軟，瀝水，去蒂切條）・・	2 克
雞髀菇（沖淨，切條）・・・・	10 克
上海冷麵・・・・・・・・	30 克

調味料

炸蒜 ・・・・・・・・・	1/2 茶匙
生抽 ・・・・・・・・・	1 茶匙
鹽 ・・・・・・・・・・	少許
糖 ・・・・・・・・・・	少許
麻油 ・・・・・・・・・	適量

做法

1 燜燒罐內注入熱水預熱，加蓋，略搖晃，靜置 1 分鐘，倒水。

2 放入蝦仁，注入熱水蓋過表面，加蓋，略搖晃，靜置 3 分鐘，瀝去水分。

3 再加入椰菜、木耳、雞髀菇和上海冷麵，注入熱水至罐內八分滿，拌勻後加蓋，略搖晃，靜置燜焗 5 分鐘，瀝去水分。

4 最後，加入調味料拌勻，完成。

鹽麴雜菜粉絲

味道清雅的雜菜粉絲,建議剪碎後才給
幼兒食用。

JCU-500-P

 材料

南瓜 (洗淨,去皮,切幼條) ‧ ‧	20 克
翠玉瓜 (洗淨,切幼條) ‧ ‧ ‧	30 克
紅甜椒 (洗淨,切幼條) ‧ ‧ ‧	20 克
黃甜椒 (洗淨,切幼條) ‧ ‧ ‧	10 克
粉絲 (浸軟,瀝水) ‧ ‧ ‧ ‧ ‧	30 克
鹽麴 ‧ ‧ ‧ ‧ ‧ ‧ ‧ ‧ ‧ ‧ ‧ ‧	1 湯匙

做法

1 燜燒罐內注入熱水預熱,加蓋,略搖晃,靜置 1 分鐘,倒水。

2 放入南瓜,注入熱水蓋過表面,拌勻後加蓋,略搖晃,靜置 8 分鐘,瀝去水分。

3 再加入翠玉瓜、紅甜椒、黃甜椒和粉絲,注入熱水至罐內八分滿,拌勻後加蓋,略搖晃,靜置燜焗 5 分鐘,瀝去水分。

4 最後,加入鹽麴拌勻,完成。

特濃茄香長通粉

採用燜燒罐烹調小朋友喜愛的西式菜式,不單方便食用,而且做法簡單,連小朋友都可一同參與煮食過程。

JBJ-302(P-P)

 材料

長通粉 · · · · · · · · ·	20 克
車厘茄(洗淨,切細) · · · ·	3 粒
黑橄欖(切片) · · · · · ·	3 粒
菠蘿(切細) · · · · · · ·	1/2 片
香草碎 · · · · · · · ·	適量

調味料

茄膏 · · · · · · · · · ·	1 湯匙
鹽 · · · · · · · · · · ·	少許
糖 · · · · · · · · · · ·	少許

做法

1 燜燒罐內注入熱水預熱,加蓋,略搖晃,靜置 1 分鐘,倒水。

2 放入長通粉和少許鹽,注入熱水蓋過表面,加蓋,略搖晃,靜置 8 分鐘,瀝去水分。

3 再次注入熱水蓋過表面,拌勻後加蓋,略搖晃,靜置 5 分鐘,瀝去水份。

4 加入車厘茄、黑橄欖和菠蘿,注入熱水至罐內八分滿,拌勻後加蓋,略搖晃,靜置燜焗 3 分鐘,瀝去水分。

5 最後,加入調味料和香草碎拌勻,完成。

海鮮酸辣湯米線

特別加入的烏醋令米線更開胃,適合夏天食用,
更可以蔬菜取代肉類,成為清爽的素湯米線。

JBM-500-G

 材料

蝦仁（沖淨） ‧ ‧ ‧ ‧ ‧ ‧ ‧	30 克
魷魚（洗淨,切細） ‧ ‧ ‧ ‧	20 克
蜆肉（沖淨） ‧ ‧ ‧ ‧ ‧ ‧	10 克
米線 ‧ ‧ ‧ ‧ ‧ ‧ ‧ ‧	100 克
番茄（去蒂,切粒） ‧ ‧ ‧ ‧	1/2 個

調味料

烏醋 ‧ ‧ ‧ ‧ ‧ ‧ ‧ ‧ ‧	1 湯匙
辣椒醬 ‧ ‧ ‧ ‧ ‧ ‧ ‧ ‧	1/2 湯匙
鹽 ‧ ‧ ‧ ‧ ‧ ‧ ‧ ‧ ‧ ‧	少許
糖 ‧ ‧ ‧ ‧ ‧ ‧ ‧ ‧ ‧ ‧	少許

做法

1. 燜燒罐內注入熱水預熱,加蓋,略搖晃,靜置 1 分鐘,倒水。

2. 放入米線,注入熱水蓋過表面,拌勻後加蓋,略搖晃,靜置 8 分鐘,瀝去水分。

3. 再放入蝦仁、魷魚和蜆肉,注入熱水蓋過食材表面,拌勻後加蓋,略搖晃,靜置 3 分鐘,瀝去水分。

4. 最後,加入番茄和調味料,注入熱水至罐內八分滿,拌勻後加蓋,略搖晃,靜置燜焗 1 分鐘,完成。

沙嗲雞絲撈麵

使用燜燒罐煮食，除了常見的烏冬外，生麵亦
是不錯的選擇。記得加入沙嗲醬，令菜式更開
胃、吸引。

SK3000MGD

 材料

雞肉（洗淨，切幼條）	50 克
甘筍	10 克
銀芽	20 克
生麵	55 克
青瓜（洗淨，切幼條） . . .	10 克
蔥（沖淨，切絲）	少許

調味料

沙嗲醬	1 湯匙
麻油	少許
鹽	適量

做法

1. 燜燒罐內注入熱水預熱，加蓋，略搖晃，靜置 1 分鐘，倒水。

2. 放入雞肉，注入熱水蓋過表面，拌勻後加蓋，略搖晃，靜置 3 分鐘，瀝去水分。

3. 再加入甘筍、銀芽和生麵，注入熱水至罐內八分滿，拌勻後加蓋，略搖晃，靜置燜焗 5 分鐘，瀝去水分。

4. 最後，加入青瓜、蔥和調味料拌勻，完成。

日式咖喱豚肉片烏冬

咖喱磚不但容易儲存,更可輕易做出香濃的咖喱的菜式。

SK3000CR

 材料

豚肉片（切細） · · · · · · ·	60 克
椰菜（洗淨，切細） · · · · ·	30 克
黃甜椒（洗淨，切幼條） · · ·	10 克
紅甜椒（洗淨，切幼條） · · ·	10 克
烏冬 · · · · · · · · · ·	1/2 個

調味料

咖喱磚（約 20 克） · · · · ·	1/6 塊
鹽 · · · · · · · · · · · ·	少許
糖 · · · · · · · · · · · ·	少許

做法

1. 燜燒罐內注入熱水預熱,加蓋,略搖晃,靜置 1 分鐘,倒水。

2. 放入豚肉片,注入熱水蓋過表面,拌勻後加蓋,略搖晃,靜置 3 分鐘,瀝去水分。

3. 再加入椰菜、黃甜椒、紅甜椒和烏冬,注入熱水至罐內八分滿,拌勻後加蓋,略搖晃,靜置燜焗 5 分鐘,瀝去水分。

4. 最後,加入調味料並注入適量水拌勻,完成。

越式魚片撈檬

保溫力佳的燜燒罐，除了熱食外，亦適合製作
冰凍的菜式於夏天隨時享用。

JBJ-302 (P-Y)

 材料

魚片（切片） · · · · · · ·	25 克
檬粉 · · · · · · ·	50 克
銀芽 · · · · · · ·	10 克
甘筍（洗淨，去皮，切幼條） · ·	10 克
西生菜（沖淨，切幼條） · · ·	15 克
羅勒（切絲） · · · · · · ·	適量
花生碎 · · · · · · ·	適量

調味料

魚露 · · · · · · ·	1 茶匙
白醋 · · · · · · ·	1 茶匙
蒜蓉 · · · · · · ·	1/2 茶匙
蜜糖 · · · · · · ·	1/2 茶匙

做法

1 燜燒罐內注入熱水預熱，加蓋，略搖晃，靜置 1 分鐘，倒水。

2 放入魚片、檬粉和銀芽，注入熱水至罐內八分滿，拌勻後加蓋，略搖晃，
靜置燜焗 3 分鐘，瀝去水分，倒出食材，放涼。

3 注入冰水冷卻燜燒罐，加蓋，略搖晃，靜置 1 分鐘，倒去冰水。

4 加入已放涼的食材、甘筍和西生菜，拌勻後加蓋，靜置保冷。

5 食用前加入羅勒、花生碎和調味料拌勻，完成。

Food Jar

Chapter 4

香焗米飯

結合了多款原汁原味、充滿食材鮮味的營
養飯餐，還額外加入兩款適合幼童的粥
品，兼顧了男女老幼的需要。

牛肝菌野菜焗紅米飯

加入牛肝菌的紅米飯，香氣會更誘人吸引。

JBM-500-G

 材料

紫洋蔥（去皮，切條）‧‧‧	20 克
蘆筍（沖淨，切段）‧‧‧‧	10 克
甘筍（去皮，切粒）‧‧‧‧	10 克
紅米‧‧‧‧‧‧‧‧‧	20 克
白米‧‧‧‧‧‧‧‧‧	60 克
牛肝菌（浸軟，瀝水，切細）‧‧‧	5 克
鹽‧‧‧‧‧‧‧‧‧‧	適量
牛油‧‧‧‧‧‧‧‧‧	適量

 做法

1 燜燒罐內注入熱水預熱，加蓋，略搖晃，靜置 1 分鐘，倒水。

2 放入紫洋蔥、蘆筍和甘筍，注入熱水蓋過食材表面，拌勻後加蓋，略搖晃，靜置 5 分鐘，瀝去水分。

3 再加入紅米和白米，注入熱水至罐內八分滿，拌勻後加蓋，略搖晃，靜置 30 分鐘，瀝去水分。

4 最後，加入牛肝菌和鹽，注入熱水至罐內八分滿，拌勻後加蓋，略搖晃，靜置燜焗 90 分鐘，食用前加入牛油拌勻，完成。

Tips

牛肝菌的獨特香氣於焗煮時能充份滲入米飯中，令香氣不易流失。

咖喱牛柳粒粟米拌飯

味道濃郁的咖喱醬,能融合牛肉和蔬菜的味道,使味道更一致。

SK3000MGD

 材料

牛柳（切粒） · · · · · · · · ·	30 克
粟米粒 · · · · · · · · ·	10 克
甘筍（去皮,切粒） · · · · ·	10 克
白米 · · · · · · · · ·	80 克
咖喱醬 · · · · · · · · ·	1 茶匙
鹽 · · · · · · · · ·	適量

 做法

1 燜燒罐內注入熱水預熱,加蓋,略搖晃,靜置 1 分鐘,倒水。

2 放入牛柳,注入熱水蓋過表面,拌勻後加蓋,略搖晃,靜置 5 分鐘,瀝去水分。

3 再加入粟米粒、甘筍和白米,注入熱水至罐內八分滿,拌勻後加蓋,略搖晃,靜置 30 分鐘,瀝去水分。

4 最後,加入咖喱醬和鹽,注入熱水至罐內八分滿,拌勻後加蓋,略搖晃,靜置燜焗 90 分鐘,完成。

Tips

改用辣度和鹹度都較低的日式兒童咖喱磚,可搖身一變成為小童都適合的咖喱飯。

櫻花蝦五香芋頭飯

除了味道非常配搭的五香粉和芋頭外,特別加入蝦香味濃郁的櫻花蝦,令鮮味更突出。

SK3000CR

 材料

芋頭(去皮,切粒) · · · · ·	50 克
白米 · · · · · · · · ·	80 克
櫻花蝦 · · · · · · · ·	3 克
五香粉 · · · · · · ·	1/2 茶匙
鹽 · · · · · · · · · ·	適量
紫菜絲 · · · · · · · ·	適量

做法

1 燜燒罐內注入熱水預熱,加蓋,略搖晃,靜置 1 分鐘,倒水。

2 放入芋頭,注入熱水蓋過表面,加蓋,略搖晃,靜置 5 分鐘,瀝去水分。

3 再加入白米,注入熱水至罐內八分滿,拌勻後加蓋,略搖晃,靜置 30 分鐘,瀝去水分。

4 最後,下櫻花蝦、五香粉和鹽,注入熱水至罐內八分滿,拌勻後加蓋,略搖晃,靜置燜焗 90 分鐘,食用前灑上紫菜絲,完成。

祛濕紅豆飯

將常見的去水腫恩物「紅豆」和「薏米」製成
正餐，能輕易地飽肚同時去濕。

JBJ-302(P-P)

 材料

紅豆（浸水一晚，瀝水）・・・	10 克
薏米（浸水一晚，瀝水）・・・	10 克
白米 ・・・・・・・・	50 克

 做法

1 燜燒罐內注入熱水預熱，加蓋，略搖晃，靜置 1 分鐘，倒水。

2 放入紅豆和薏米，注入熱水蓋過食材表面，拌勻後加蓋，略搖晃，靜置 30 分鐘，瀝去水分。

3 加入白米，注入熱水至罐內八分滿，拌勻後加蓋，略搖晃，靜置 30 分鐘，瀝去水分。

4 再次注入熱水至罐內八分滿，拌勻後加蓋，略搖晃，靜置燜焗 90 分鐘，完成。

韓式辣醬魷魚拌飯

一道香辣惹味的韓式拌飯，除了魷魚外，
採用其他海鮮亦可。

JBJ-302 (P-Y)

 材料

魷魚（沖淨，切件）	30 克
甘筍（去皮，切條）	10 克
豆芽	10 克
海帶（浸軟，瀝水，切條）	少許
白米	50 克
韓式辣醬	1 茶匙
生抽	1/2 茶匙
麻油	適量
蛋絲	10 克

做法

1 燜燒罐內注入熱水預熱，加蓋，略搖晃，靜置 1 分鐘，倒水。

2 放入魷魚，注入熱水蓋過表面，加蓋，略搖晃，靜置 5 分鐘，瀝去水分。

3 再加入甘筍、豆芽、海帶和白米，注入熱水至罐內八分滿，拌勻後加蓋，
略搖晃，靜置 30 分鐘，瀝去水分。

4 再次注入熱水至罐內八分滿，拌勻後加蓋，略搖晃，靜置燜焗 90 分鐘，
食用前拌入韓式辣醬、生抽、麻油和蛋絲，完成。

黑蒜雞粒焗飯

經過焦糖化的黑蒜帶有香甜味道，為焗飯增加
天然甜味。

SK3020BK

 材料

雞肉（切粒） · · · · · · · · · · · · ·	50 克
洋蔥（去皮，切粒） · · · · · · · · · ·	20 克
紅、黃、青甜椒（洗淨，切粒） · · · · · ·	各 10 克
白米 · · · · · · · · · · · · · · ·	100 克
黑蒜（切粒） · · · · · · · · · · · ·	4 粒
鹽 · · · · · · · · · · · · · · · ·	適量

做法

1 燜燒罐內注入熱水預熱，加蓋，略搖晃，靜置 1 分鐘，倒水。

2 放入雞肉，注入熱水蓋過表面，拌勻後加蓋，略搖晃，靜置 5 分鐘，瀝去水分。

3 再加入洋蔥、紅、黃、青甜椒和白米，注入熱水至罐內八分滿，拌勻後加蓋，略搖晃，靜置 30 分鐘，瀝去水分。

4 最後，下黑蒜和鹽，注入熱水至罐內八分滿，拌勻後加蓋，略搖晃，靜置燜焗 90 分鐘，完成。

雙色番薯粥

使用燜燒罐「無火」煮食，簡單又方便地將高纖維、營養豐富的番薯煮熟，非常適合小朋友進食。

JBM-500-GRP

 材料

黃芯番薯（去皮，切粒）...	20 克
紫芯番薯（去皮，切粒）...	20 克
白米	50 克

 做法

1 燜燒罐內注入熱水預熱，加蓋，略搖晃，靜置 1 分鐘，倒水。

2 放入黃芯番薯和紫芯番薯，注入熱水蓋過表面，加蓋，略搖晃，靜置 5 分鐘，瀝去水分。

3 加入白米，注入熱水至罐內八分滿，拌勻後加蓋，略搖晃，靜置 30 分鐘，瀝去水分。

4 再次注入熱水至罐內八分滿，拌勻後加蓋，略搖晃，靜置燜焗 90 分鐘，完成。

大良牛乳燕麥粥

加入帶有淡淡鹹香,有「中式芝士」之稱的大
良牛乳,為香綿白粥加添滋味。

SK3020WH

 材料

白米 · · · · · · · · ·	1/2 量米杯
即食燕麥 · · · · · · · ·	1 湯匙
白果（去殼,切半,去芯）· · ·	10 克
大良牛乳 · · · · · · · ·	2 片

做法

1 燜燒罐內注入熱水預熱,加蓋,略搖晃,靜置 1 分鐘,倒水。

2 放入白果,注入熱水蓋過表面,加蓋,略搖晃,靜置 5 分鐘,瀝去水分。

3 加入即食燕麥和白米,注入熱水至罐內八分滿,拌勻後加蓋,略搖晃,
靜置 30 分鐘,瀝去水分。

4 再次注入熱水至罐內八分滿,拌勻後加蓋,略搖晃,靜置燜焗 90 分鐘,
食用前加入大良牛乳,完成。

Shuttle Chef

Chapter 5
窩心湯品

適合忙碌都市人的滋補湯品，不用長期睇火，既安全又方便，「煲」出的湯水能夠媲美足料老火湯的味道和營養。

芥菜豆腐鮮魚湯

簡單易做的清熱魚湯，最適合工作忙碌、無暇
做飯的香港人。

RPE-3000CA

 材料 (3-5人用)

薑片 · · · · · · · · ·	2 片
蔥段 · · · · · · · · ·	適量
鯇魚尾 (洗淨，用廚紙印乾) · ·	1 條
紹興酒 · · · · · · · · ·	適量
熱水 · · · · · · · · ·	4 杯
芥菜 (沖淨，切件) · · · ·	300 克
豆芽 · · · · · · · · ·	30 克
豆腐 (切件) · · · · ·	360 克
鹽 · · · · · · · · · ·	適量

 做法

1 真空煲內鍋下油加熱，爆香薑片和蔥段，下鯇魚尾煎香。

2 灒入紹興酒，注入熱水。

3 下芥菜、豆芽和豆腐，煮滾後加蓋，以中火煮 5 分鐘，熄火，放入真空
煲外鍋燜焗 30 分鐘，加鹽調味，完成。

Tips

將魚洗淨後放入鍋內煎至金黃後，注入熱水煮滾，就可以令湯色變白。

椰香海底椰百合湯

海底椰的果肉潔白可口，氣味清香，有潤肺止咳，清熱之功效，非常適合小朋友飲用，秋天時份甚佳。

RPC-6000(6L)

 材料 (6-8人用)

瘦肉 · · · · · · · · · · · · · ·	100	克
新鮮海底椰 (去皮，沖淨，切半) · · · ·	100	克
乾百合 (沖淨，瀝水) · · · · · · ·	10	克
甘筍 (去皮，洗淨，切件) · · · · · · ·	1	條
粟米 (去衣，洗淨，切件) · · · · ·	1	條
蘋果 (切件，去籽) · · · · · ·	1	個
蜜棗 · · · · · · · · · · · ·	2	粒
水 · · · · · · · · · · · · · ·	6	杯

做法

1 瘦肉汆水，盛起瀝水。

2 真空煲內鍋放入所有食材，煮滾後加蓋，以中火煮 5 分鐘，熄火，放入真空煲外鍋燜焗 2 小時，完成。

紅菜頭腰果素湯

要烹調一道美味的素湯，秘訣在於要添加足量
的腰果及乾冬菇等食材，特別是豐富的果仁味
能令素湯味道香濃、飽肚又不油膩。

RPE-3000OLV

 材料 (3-5人用)

紅菜頭 （去皮，切件）・・・・	200 克
甘筍 （去皮，切件）・・・・	100 克
栗子 （去殼，去衣）・・・・	130 克
粟米 （去衣，洗淨，切件）・・・	1 條
雪梨 （去皮，切件，去籽）・・・	1 個
腰果 ・・・・・・・	60 克
乾冬菇 ・・・・・・・	2 隻
雪耳 （浸軟，瀝水，去蒂）・・	20 克
蜜棗 ・・・・・・・・	2 粒
水 ・・・・・・・・・・・	6 杯

做法

1 所有材料放入真空煲內鍋，煮滾後加蓋，以中火煮 5 分鐘，熄火，放入
真空煲外鍋燜焗 2 小時，完成。

蟲草花淮蓮螺頭湯

蟲草花不寒不燥,有滋陰補腎、抗衰老、提高免疫力的功效。利用真空煲,只需烹煮 5 分鐘,然後放在外鍋內保溫燜焗 2 小時,完成品的味道便與長時間煲煮的老火湯一樣突出。

KPS-8000

 材料(10-12 人用)

蟲草花(浸軟,瀝水)	20 克
乾淮山(沖淨)	40 克
蓮子(沖淨,去芯)	30 克
茨實(沖淨)	30 克
急凍螺頭(沖淨)	400 克
瘦肉	250 克
水	10 杯
鹽	適量

做法

1 急凍螺頭和瘦肉汆水,盛起瀝水。

2 真空煲內鍋放入所有食材(除鹽外),煮滾後加蓋,以中火煮 5 分鐘,熄火,放入真空煲外鍋燜煮 2 小時,加鹽調味,完成。

羅漢果粟米鮑魚骨湯

鮑魚骨即是豬的腮骨，肉質口感像牛展。煲湯後，肉質還是那麼軟綿，是我煲湯首選材料之一。

KBA-4501-TOM

 材料 (4-6人用)

羅漢果（略壓碎）	. . . . . .	1 個
粟米（去衣，洗淨，切件）	. . .	1 條
甘筍（去皮，切件）	. . . . . .	1 條
鮑魚骨	. . . .	300 克
馬蹄（去皮）	. . . . . .	8 粒
蜜棗	. . . .	2 粒
水	. . . .	6 杯
鹽	. . . .	適量

做法

1 鮑魚骨汆水，盛起沖淨，瀝水。

2 真空煲內鍋放入所有食材（除鹽外），煮滾後加蓋，以中火煮 5 分鐘，熄火，放入真空煲外鍋燜焗 2 小時，加鹽調味，完成。

Shuttle Chef

Chapter 6

滋味燜餸

真空煲節能又省電,只需花短時間燜煮,
已令所有食材非常入味又軟腍,遇上家
人、朋友聚餐都不用煩惱。

潮式滷水牛腱

使用多種不同香料浸煮而成的滷水牛腱，充當
熱食、冷盤都一樣滋味。

RPC-6000(6L)

材料（6-8人用）

急凍牛腱（用刀切去筋膜） · · · · 1000 克

香料

香葉 · · · · · · · · · · ·	5 塊
八角 · · · · · · · · ·	10 克
玉桂皮 · · · · · · · ·	2 克
白胡椒粒 · · · · · · ·	1 茶匙
丁香 · · · · · · · · · ·	少許
沙薑 · · · · · · · · · ·	2 塊
草果 · · · · · · · · · ·	4 粒
甘草 · · · · · · · · · ·	5 克

汁料

蒜頭（去皮） · · · · ·	3 粒
薑 · · · · · · · · · ·	3 片
乾蔥（去皮，切件） · ·	2 粒
紹興酒 · · · · · · · ·	適量
冰糖 · · · · · · · · ·	30 克
蠔油 · · · · · · · · ·	1 湯匙
生抽 · · · · · · · · ·	3 湯匙
老抽 · · · · · · · · ·	1 湯匙
鹽 · · · · · · · · · ·	1 茶匙
水 · · · · · · · · · ·	6 杯

 做法

1 所有香料放入煲湯袋內，束緊後放入真空煲內鍋，下牛腱，注入汁料煮滾，
加蓋，以中火煮 20 分鐘。

2 熄火，放入真空煲外鍋燜焗 90 分鐘，取出牛腱切片，完成。

Tips

此做法的牛腱較有咬口，如喜歡肉質更軟、味道更濃的話，可將牛腱取出切片後，
放回真空煲內鍋以中火煮滾後加蓋，續煮 5 分鐘，再放入真空煲外鍋燜焗 30 分
鐘即可。

杞子話梅炆雞翼

炆煮的雞翼比煎製的雞翼較少油膩感，配以酸甜的話梅，味道更加開胃，伴飯一流。

RPE-3000CA

 材料 (3-5人用)

雞翼 · · · · · · · · · · · · ·	10 隻
蒜頭（拍扁，去皮）· · · · · · ·	2 粒
紫洋蔥（去皮，切件）· · · · · ·	1/2 個
蘋果（切件，去籽）· · · · · · ·	1/2 個
話梅 · · · · · · · · · · · · ·	5 粒
冰糖 · · · · · · · · · · · · ·	適量
老抽 · · · · · · · · · · · · ·	1 湯匙
生抽 · · · · · · · · · · · · ·	2 湯匙
水 · · · · · · · · · · · · · ·	300 毫升
杞子 （浸軟，瀝水）· · · · · · ·	1 湯匙

做法

1 真空煲內鍋下油加熱，下雞翼煎至金黃，盛起。

2 爆香蒜頭和紫洋蔥，雞翼回鑊，加入蘋果、話梅、冰糖、老抽和生抽，注入水以中火煮滾，加蓋，以中火續煮 5 分鐘。

3 下杞子拌勻，熄火，放入真空煲外鍋爛焗 30 分鐘，完成。

韓醬蓮藕炆豬軟骨

豬軟骨經過真空煲燜煮後變得軟稔，但仍保持一定的口感，配上蓮藕後比一般韓式豬軟骨更豐富。

KBA-4501-TOM

 材料 (4-6人用)

豬軟骨 · · · · · · · ·	300 克
韓式醃肉汁 · · · · · ·	2 湯匙
洋蔥 (去皮，切件) · · · ·	1/2 個
甘筍 (去皮，切件) · · · ·	120 克
蓮藕 (去皮，切件) · · · ·	350 克
水 · · · · · · · · ·	350 毫升

調味料

蒜蓉 · · · · · · · ·	1 茶匙
韓式辣椒醬 · · · · · · ·	1 湯匙
白醋 · · · · · · · · ·	1 湯匙
糖 · · · · · · · · ·	1 茶匙

做法

1 碗內放入豬軟骨和韓式醃肉汁拌勻。

2 真空煲內鍋下油加熱，下豬軟骨以中慢火煎炒至金黃，盛起。

3 再加入洋蔥和甘筍炒香，下蓮藕略炒，豬軟骨回鑊，加入調味料拌勻，注入水以中火煮滾，加蓋續煮 10 分鐘，熄火，放入真空煲外鍋燜焗 90 分鐘，完成。

五香枝竹牛腩煲

一般燜煮牛腩需要利用大量能源長時間燜煮，但採用真空煲烹煮，便能夠節能減碳，提高便利度。這款惹味菜式最適合冬天時節食用。

KBA-4501-TOM

材料 (4-6人用)

牛腩（切件）	· · · ·	1000 克
薑片	· · · · ·	3 片
蔥（切段）	· · · ·	適量
水	· · · ·	500 毫升
白蘿蔔	· · · ·	600 克
紅尖椒	· · · · ·	1 條
枝竹	· · · · ·	200 克
豆卜	· · · · ·	8 件

調味料

五香粉	· · · ·	1/2 茶匙
鹽	· · · · ·	適量
黑胡椒粉	· · · ·	適量
柱侯醬	· · · ·	3 湯匙
蠔油	· · · ·	2 湯匙
冰糖	· · · ·	適量

做法

1 碗內放入牛腩、五香粉、鹽和黑胡椒粉拌勻。

2 真空煲內鍋下油加熱，爆香薑片和蔥，下牛腩以中火煎炒至金黃，加入柱侯醬略炒。

3 注入水，加入蠔油、冰糖、白蘿蔔和紅尖椒煮滾，加蓋，以中火續煮 20 分鐘。

4 下枝竹和豆卜拌勻，熄火，放入真空煲外鍋燜焗 90 分鐘，完成。

香菇淮山栗子炆雞

利用真空煲短時間烹煮，長時間燜焗的功能，令菜式在焗煮的過程中不會有蒸氣流失，使食材的香氣及營養凝聚於煲內，有效地保存食物的原汁原味。

RPE-3000CA

 材料 (3-5人用)

栗子	100 克
雞件 (洗淨，瀝水)	400 克
乾蔥 (去皮，切件)	2 粒
蒜頭 (拍扁，去皮)	2 粒
甘筍 (去皮，切件)	120 克
乾冬菇 (浸軟，瀝水，去蒂切片)	30 克
鮮淮山 (去皮，切件) . . .	150 克
西蘭花 (切小朵，沖淨) . . .	100 克
水	300 毫升

 醃料

中式麵豉醬	1 湯匙
米酒	1 茶匙
麻油	適量

做法

1 栗子放入小煲內，注入水蓋過栗子，以中火煮滾，熄火，加入一半冷水降溫，隨即以小刀去殼、去衣，備用。

2 雞件和醃料放碗內拌勻，備用。

3 真空煲內鍋下油加熱，爆香乾蔥和蒜頭，加入雞件煎炒至金黃，下甘筍炒香。

4 加入栗子、乾冬菇、鮮淮山和西蘭花拌勻，注入水煮滾，加蓋後以中火續煮 5 分鐘，熄火，放入真空煲外鍋燜焗 45 分鐘，完成。

豉油牛肝菌滑雞

帶有淡淡菇菌香氣的豉油雞，味道較傳統的清香，適合注重健康養身的現代人。

KPS-8000

 材料（10-12人用）

雞（沖淨）	1 隻
薑	2 片
乾蔥（去皮，切件）	3 粒
甘筍（去皮，切件）	60 克
牛肝菌（浸軟，瀝水）	20 克
乾冬菇（浸軟，瀝水，去蒂切片）	50 克
蟹爪菇（切去根部，沖淨）	50 克

醃料

玫瑰露酒	1 茶匙
老抽	1 湯匙
生抽	2 湯匙

汁料

生抽	60 毫升
老抽	2 湯匙
冰糖	50 克
香葉	2 片
水	14 杯

做法

1 醃料混合，塗抹於雞隻表面，醃 15 分鐘。

2 真空煲內鍋下油加熱，爆香薑、乾蔥和甘筍，下牛肝菌、乾冬菇和蟹爪菇炒香。

3 放入雞（連剩餘醃料），注入汁料煮滾，加蓋以中火續煮 10 分鐘，熄火，放入真空煲外鍋燜焗 60 分鐘，完成。

Shuttle Chef

Chapter 7
新穎粥品糖水

只需騰出早上少許時間煮開米粒,不需一
小時,不用睇火便可輕易熬出香綿滑粥。
另外,我亦十分推薦以下的三款清潤糖
水,加入養生食材,將傳統糖水變作更適
合現代人的口味。

桂花蘋果雪梨雪蓮子

雪蓮子是美顏食品，含豐富膠質，有「植物界燕窩」之稱，加入後能令香甜的糖水更滋潤。

RPC-6000(6L)

 材料（6-8人用）

桂花乾 · · · · · · · · ·	1 湯匙
雪蓮子（浸水一晚，瀝水） · ·	60 克
蘋果（洗淨，切件，去籽） · · · ·	2 個
雪梨（洗淨，切件，去籽） · · · ·	2 個
水 · · · · · · · · ·	12 杯
冰糖 · · · · · · · · ·	60 克

做法

1 所有材料放入真空煲內鍋（除冰糖外），加蓋煮滾，以中火續煮 5 分鐘。

2 下冰糖拌煮至溶，加蓋以中火再次煮滾，熄火，放入真空煲外鍋燜焗 30 分鐘，完成。

腐竹薏米桃膠糖水

桃膠是桃樹皮中分泌出來的樹脂，蘊含豐富植物膠原，性質溫和，質感煙韌像蒟蒻，為此款既清熱又滋潤的糖水增添口感。

RPE-3000CA

 材料 (3-5人用)

腐竹 · · · · · · · · ·	15 克
桃膠 （浸水一晚，洗淨，瀝水）·	40 克
冰糖 · · · · · · · · ·	90 克
薏米 · · · · · · · · ·	30 克
水 · · · · · · · · ·	6 杯

做法

1 所有材料放入真空煲內鍋（除冰糖外），加蓋煮滾，以中火續煮 10 分鐘。

2 下冰糖拌煮至溶，加蓋後以中火再次煮滾，熄火，放入真空煲外鍋燜焗 90 分鐘，完成。

蛋花紫薯年糕糖水

煙韌的年糕配上粉嫩的紫薯，色相和味道都份
外吸引。

KBA-4501-TOM

 材料（4-6人用）

紫薯（去皮，切件） · · · · 320 克

年糕 · · · · · · · · · 150 克

冰糖 · · · · · · · · · 140 克

水 · · · · · · · · · · 8 杯

雞蛋 · · · · · · · · · 1 隻

做法

1 真空煲內鍋注入水煮滾，下紫薯、年糕和冰糖拌勻，加蓋煮滾，以中火
煮 3 分鐘。

2 放入真空煲外鍋燜焗 15 分鐘，開蓋，加入已打散的雞蛋，拌勻成蛋花，
完成。

無花果雪梨百合粥

性質溫和又滋潤的甜粥,適合任何體質人仕食用。

RPE-3000OLV

 材料 (3-5人用)

無花果乾 (切半) · · · · · · 2 件

雪梨 (去皮,切件,去籽) · · · 1 個

白米 · · · · · · · 3/4 個量米杯

水 · · · · · · · · · · 6 杯

鮮百合 (沖淨,分成小瓣) · · · 2 個

杞子 (浸軟,瀝水) · · · · 1 湯匙

冰糖 · · · · · · · · · 30 克

做法

1 無花果乾、雪梨、白米和水放入真空煲內鍋,加蓋煮滾,以中火續煮 5 分鐘。

2 下鮮百合、杞子和冰糖拌勻,加蓋以中火再次煮滾,熄火,放入真空煲外鍋燜焗 30 分鐘,完成。

番茄大蝦五穀粥

將材料先炒後煮可以令這款菜式的味道更加豐富及惹味。

RPE-3000OLV

 材料 (3-5人用)

蒜蓉 · · · · · · · · ·	1 茶匙
洋蔥（去皮，切粒） · · · ·	1/2 個
鮮蝦 · · · · · · · · ·	6 隻
番茄（去蒂，切粒） · · · · ·	2 個
茄膏 · · · · · · · · ·	2 湯匙
鹽 · · · · · · · · · ·	1 茶匙
糖 · · · · · · · · · ·	1 湯匙
白米 · · · · · · ·	1.5 量米杯
燕麥 · · · · · · · · ·	2 湯匙
三色藜麥 · · · · · · ·	1 湯匙
水 · · · · · · · · · ·	6 杯

做法

1 真空煲內鍋下油加熱，爆香蒜蓉和洋蔥，下鮮蝦炒熟。

2 加入番茄和茄膏拌勻，下鹽和糖調味，下白米、燕麥和三色藜麥拌勻，注入水煮滾，加蓋，以中火續煮 5 分鐘，熄火，放入真空煲外鍋燜焗 30 分鐘，完成。

Appendix

食譜英文翻譯

需要書中食譜的英文版本？本章提供了全
書 45 道食譜的全英文翻譯，方便了外籍
人士閱讀及取用。

Pear, Red Dates and Chestnut Sweet Soup

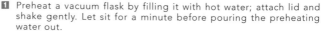

P.18

 INGREDIENTS

1/2 pear, peeled, seeded and diced

2 red dates, soaked to soften, pitted and cut into small pieces

2 water chestnuts, peeled and diced

1 tablespoon water chestnut flour

1 1/2 tablespoon sugar

 METHOD

1 Preheat a vacuum flask by filling it with hot water; attach lid and shake gently. Let sit for a minute before pouring the preheating water out.

2 Put the pear, red dates and water chestnuts into the flask. Fill it with boiling water to cover the ingredients. Stir gently, attach lid and shake lightly. Let sit for 15 minutes before draining.

3 Add water chestnut flour and sugar to the flask. Pour in boiling water until the flask is 80% full, stir quickly and thoroughly. Attach lid and shake gently. Let sit for 5 minutes before serving.

TIPS Pears not only add sweetness to the soup, but also clear heat according to Chinese medicine principles; feel free to substitute with apples if preferred. The water chestnut flour also adds a lovely thickness to the soup.

Ginger, Red Beans and Taro Sweet Soup

P.20

 INGREDIENTS

30g taro, peeled and diced

A dash Japanese sweetened red beans

1 tablespoon ginger juice

 METHOD

1 Preheat a vacuum flask by filling it with hot water; attach lid and shake gently. Let it sit for a minute before pouring the preheating water out.

2 Add diced taro to the flask, fill it with boiling water to cover the taro. Stir gently, attach lid and shake slightly. Let sit for 5 minutes before pouring the water out.

3 Put red beans and ginger juice into the flask. Fill the flask with water until it is 80% full, stir gently. Attach lid and shake lightly. Let rest for 5 minutes before serving.

TIPS The dessert is extremely easy to make with the use of canned Japanese sweetened red beans; while the ginger juice makes it the perfect stomach-warming sweet soup for winter.

Konnyaku Noodles with Strawberry Sauce

P.22

 INGREDIENTS

80g strawberries, hulled, rinsed and diced

1 tablespoon honey

1 ice cube

120g konjac noodles

 TIPS

Konjac noodles can be served both savory and sweet. You can also chill it in the fridge and serve cold with strawberry sauce.

 METHOD

1 Blend 60g strawberries, honey and ice cube in a blender. Transfer the mixture to a bowl and set aside.

2 Preheat a vacuum flask by filling it with hot water; attach lid and shake gently. Let it sit for a minute before pouring the preheating water out.

3 Add noodles to the flask and fill with boiling water until the flask is 80% full. Stir gently and attach lid. Shake slightly then let sit for 3 minutes. Drain the noodles and let cool.

4 Cool down the flask by filling it with ice water. Attach lid and shake slightly. Let sit for a minute and pour the water out.

5 Place the noodles, strawberry sauce and the remaining strawberries into the flask. Stir gently and attach lid to keep it chilly.

Peach Gum, Corn and Papaya Sweet Soup

P.22

 INGREDIENTS

5g peach gum, soaked overnight, rinsed and drained

20g sweet corn kernels

25g papaya, skinned and diced

 METHOD

1. Blend 15g sweet corns and 100ml water in a blender. Strain the mixture, keep 50ml of the juice and set aside.

2. Preheat a vacuum flask by filling it with hot water; attach lid and shake gently. Let it sit for a minute before pouring the preheating water out.

3. Add the peach gum, sweet corn juice, diced papaya and the remaining sweet corn kernels to the flask. Fill the flask with boiling water until it is 80% full. Stir gently, shake lightly and let sit for 25 minutes before serving.

 TIPS   Soak the peach gum overnight and drain in a strainer. Rinse and remove the black impurities.

Purple Sweet Potato and Cous cous with Coconut Milk

P.26

 INGREDIENTS

35g purple sweet potatoes, peeled and diced

1 tablespoon cous cous

1 tablespoon coconut milk

1/2 tablespoon honey

 METHOD

1. Preheat a vacuum flask by filling it with hot water; attach lid and shake gently. Let it sit for a minute before pouring the preheating water out.

2. Add diced sweet potato to the flask, fill it with boiling water to cover the sweet potatoes. Stir gently, attach lid and shake slightly. Let sit for 5 minutes before draining.

3. Put cous cous into the flask. Fill the flask with boiling water until it's 80% full. Stir gently, attach lid and shake slightly. Let sit for 5 minutes before stirring in coconut milk and honey to serve.

 TIPS   Cous cous is one of the few grains that is alkalizing to the body, and it is easy to cook and digest. Therefore, millet is perfect for vacuum bottle cooking.

Pork, Zucchini and Cloud Ear Mushroom in Oyster Sauce

P.30

INGREDIENTS

50g pork slices, cut into bite-size

40g zucchini, rinsed and cut into thin slices

1g cloud ear mushrooms, soaked to soften, hard ends removed and chopped

SEASONINGS

A pinch of salt

1 tablespoon oyster sauce

METHOD

1 Preheat a vacuum flask by filling it with hot water; attach lid and shake gently. Let it sit for a minute before pouring the preheating water out.

2 Put the pork into the flask and pour in boiling water to cover the pork. Stir slightly, attach lid and shake gently. Let sit for 2 minutes and drain.

3 Add in the zucchini and mushroom pieces. Fill the flask with hot water until it is 80% full. Stir slightly, attach lid and shake gently. Let sit for 5 minutes and drain.

4 Stir in seasonings to serve.

TIPS Use a pair of chopsticks to spread the pork slices evenly in the flask, to make sure that all of them will be heated and cooked thoroughly.

Miso Braised Pork with Bitter Melon

P.32

INGREDIENTS

50g pork, rinsed and diced

70g bitter melon, rinsed and cut into thin slices

2 fresh shiitake mushrooms, rinsed, stems removed and cut into slices

Some Japanese style pickled burdock

SEASONINGS

A pinch of salt

A pinch of sugar

1/2 tablespoon miso paste

1/2 tablespoon water

METHOD

1 Preheat a vacuum flask by filling it with hot water; attach lid and shake gently. Let it sit for a minute before pouring the preheating water out.

2 Put the pork into the flask and pour in boiling water to cover the pork. Stir slightly, attach lid and shake gently. Let sit for 5 minutes and drain.

3 Add the bitter melon and mushrooms to the flask. Fill the flask with hot water until it is 80% full. Stir slightly, attach lid and shake gently. Let sit for 5 minutes and drain.

4 Mix together with the seasonings and pickled burdock to serve.

TIPS The extreme bitter taste of the bitter melon is removed after braising. Those who are not fans of bitter melon may find the dish worth trying.

French Bean and Shrimp in Black Truffle Sauce

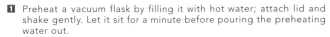

P.34

 INGREDIENTS

12 small shrimps, peeled and deveined

70g French beans, rinsed and cut into lengths

1/3 yellow bell pepper, rinsed, pitted and cut into strips

 SEASONINGS

A pinch of salt

1/2 teaspoon black truffle paste

A drizzle of olive oil

 METHOD

1. Preheat a vacuum flask by filling it with hot water; attach lid and shake gently. Let it sit for a minute before pouring the preheating water out.

2. Put the shrimps into the flask and pour in boiling water to cover the shrimps. Stir slightly, attach lid and shake gently. Let sit for a minute and drain.

3. Add the beans and pepper pieces to the flask. Fill the flask with hot water until it is 80% full. Stir slightly, attach lid and shake gently. Let sit for 3 minutes and drain.

4. Stir in the seasonings to serve.

 TIPS Shrimps are high in protein so they needed to be blanched before cooking, to make sure that the dish is free of foam.

Sliced Dover Sole and Eggplant in XO Sauce

P.36

 INGREDIENTS

80g dover sole fillet, sliced into thin strips

1/2 eggplant, rinsed and cut into strips

20g snow peas, rinsed and strings removed

 SEASONINGS

A pinch of salt

A pinch of sugar

1/2 teaspoon XO sauce

1/2 teaspoon oyster sauce

 METHOD

1. Preheat a vacuum flask by filling it with hot water; attach lid and shake gently. Let it sit for a minute before pouring the preheating water out.

2. Put the fish slices into the flask. Fill the flask with boiling water to cover the fish. Stir slightly, attach lid and shake gently. Let sit for 3 minutes and drain.

3. Add in the eggplant and peas. Fill the flask with hot water until it's 80% full. Stir slightly, attach lid and shake gently. Let sit for 5 minutes and drain.

4. Stir in the seasonings and enjoy!

 TIPS The spiciness added by XO sauce will sure drive your appetite up!

Scallop and Snap Pea with Kimchi & Orange Marmalade

 INGREDIENTS

30g baby scallops

30g snap peas, rinsed and strings removed

30g baby corns, rinsed and cut into small rounds

30g kimchi, cut into small pieces

 SEASONINGS

1 teaspoon orange marmalade

A pinch of salt

 METHOD

P.38

1. Preheat a vacuum flask by filling it with hot water; attach lid and shake gently. Let it sit for a minute before pouring the preheating water out.

2. Put scallops into the flask and pour in boiling water to cover the scallops. Stir slightly, attach lid and shake gently. Let sit for 3 minutes and drain.

3. Add in snap peas and baby corns. Fill the flask with hot water until it is 80% full. Stir slightly, attach lid and shake gently. Let sit for 3 minutes and drain.

4. Add kimchi and seasonings. Stir lightly and serve.

 TIPS For some bite, add some bite-size orange pieces in the last step to the dish before serving.

Diced Pork with Mushroom and Tofu Puff

 INGREDIENTS

65g pork, diced

20g tofu puffs, rinsed and cut into halves

2 fresh shiitake mushrooms, rinsed, stems removed and cut into halves

20g oyster mushrooms, rinsed and torn into pieces

2 baby Chinese cabbage leaves, rinsed and cut into pieces

 SEASONINGS

A pinch of sugar

1 teaspoon black bean garlic sauce

 METHOD

P.40

1. Preheat a vacuum flask by filling it with hot water; attach lid and shake gently. Let it sit for a minute before pouring the preheating water out.

2. Put pork into the flask and pour in boiling water to cover the pork. Stir slightly, attach lid and shake gently. Let sit for 5 minutes and drain.

3. Add in the tofu puffs, mushrooms and cabbage leaves. Fill the flask with hot water until it's 80% full. Stir slightly, attach lid and shake gently. Let sit for 5 minutes and drain.

4. Stir in seasonings to serve.

 TIPS Black bean garlic sauce gives the dish an authentic Hong Kong flavor.

Thai Chicken Curry with Basil

 INGREDIENTS

80g chicken, rinsed and diced

20g cauliflower, rinsed and cut into bite-size

20g broccoli, rinsed and cut into bite-size

20g carrot, rinsed, peeled and cut into thin slices

A bunch of basil leaves, chopped

 SEASONINGS

1/2 tablespoon green curry paste

1 teaspoon coconut milk

A pinch of salt

 METHOD

1 Preheat a vacuum flask by filling it with hot water; attach lid and shake gently. Let it sit for a minute before pouring the preheating water out.

2 Put the diced chicken into the flask and pour in boiling water to cover the chicken. Stir slightly, attach lid and shake gently. Let sit for 3 minutes and drain.

3 Add the cauliflower, broccoli and carrot to the flask. Fill the flask with hot water until it is 80% full. Stir slightly, attach lid and shake gently. Let sit for 5 minutes and drain.

4 Mix together with basil and seasonings to serve.

 TIPS   Basil lends a refreshing tang to the dish and makes it tastes lighter and more appetizing.

Five Spice Shrimp Ball in Tomato Sauce

P.44

 INGREDIENTS

1/2 tomato, rinsed, hulled and cut into small pieces

20g thin green asparagus, rinsed and cut into lengths

20g onion, peeled and minced

1/2 tablespoon tomato paste

A pinch of salt

A pinch of sugar

 SHRIMP BALLS

50g shrimps, peeled, deveined and cut into pieces.

A pinch of white pepper powder

A pinch of salt

A pinch of five spice powder

5g sweet corn kernels

 METHOD

1 For the shrimp balls, blend shrimps, pepper powder, salt and five spice powder in a blender. Mix with the corns in another bowl. Then shape the shrimp mixture into balls; make 5 balls.

2 Preheat a vacuum flask by filling it with hot water; attach lid and shake gently. Let it sit for a minute before pouring the preheating water out.

3 Fill the flask with hot water until it is 80% full, add the shrimp balls to the flask. Stir slightly, attach lid and shake gently. Let sit for 5 minutes and drain.

4 Add in the tomato, asparagus and onion pieces. Fill the flask with hot water until it is 80% full. Stir slightly, attach lid and shake gently. Let sit for 3 minutes and drain.

5 Stir in the tomato paste, salt and sugar to serve.

 TIPS   Remember to fill the flask with hot water before putting in the shrimp balls, so that the balls can hold their shape.

127

Udon with Preserved Olive Leaves and Minced Pork

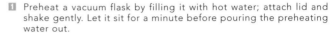

 INGREDIENTS

50g minced pork

20g sweet corn kernels

1 sting Chinese string bean, rinsed and chopped

1/2 udon

 SEASONINGS

1 tablespoon Preserved olive leaves

A pinch of salt

A pinch of sugar

 METHOD

1. Preheat a vacuum flask by filling it with hot water; attach lid and shake gently. Let it sit for a minute before pouring the preheating water out.

2. Put the minced pork into the flask. Fill the flask with boiling water to cover the pork. Stir slightly, attach lid and shake gently. Let sit for 3 minutes and drain.

3. Add in the corns, chopped bean and udon. Fill the flask with hot water until it's 80% full. Stir slightly, attach lid and shake gently. Let sit for 5 minutes and drain.

4. Stir in the seasonings to serve.

 TIPS Preserved olive leaves gives this Chiuchow-style udon a richer taste.

Noodles with Garlic and Shrimp

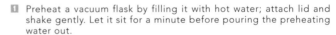

P.50

 INGREDIENTS

30g small shrimps, rinsed and deveined

20g cabbage, rinsed and shredded

2g cloud ear mushrooms, soaked to softened, hard ends removed and shredded

10g king oyster mushrooms, rinsed and shredded

30g Shanghai white noodles

 SEASONINGS

1/2 teaspoon fried garlic

1 teaspoon soy sauce

A pinch of salt

A pinch of sugar

A drizzle of sesame oil

 METHOD

1. Preheat a vacuum flask by filling it with hot water; attach lid and shake gently. Let it sit for a minute before pouring the preheating water out.

2. Put the shrimps into the flask. Fill the flask with boiling water to cover the shrimps. Stir slightly, attach lid and shake gently. Let sit for 3 minutes and drain.

3. Add in the cabbage, mushrooms and noodles. Fill the flask with hot water until it's 80% full. Stir slightly, attach lid and shake gently. Let sit for 5 minutes and drain.

4. Stir in seasonings to serve.

 TIPS This easy and simple noodle dish is tasty and rich in nutrition with all the vegetables and mushrooms.

Mixed Vegetables with Vermicelli in Shio Koji Sauce

 INGREDIENTS

20g pumpkin, rinsed, peeled and cut into thin strips

30g zucchini, rinsed and shredded

20g red bell pepper, rinsed and cut into thin strips

10g yellow bell pepper, rinsed and cut into thin strips

30g glass noodles, soaked to soften and drained

1 tablespoon shio koji (a mix of Japanese malted rice, salt and water which is a kind of seasoning)

 METHOD

1. Preheat a vacuum flask by filling it with hot water; attach lid and shake gently. Let it sit for a minute before pouring the preheating water out.

2. Put the pumpkin into the flask. Fill the flask with boiling water to cover the pumpkin. Stir slightly, attach lid and shake gently. Let sit for 8 minutes and drain.

3. Add in the zucchini, bell peppers and noodles. Fill the flask with hot water until it's 80% full. Stir slightly, attach lid and shake gently. Let sit for 5 minutes and drain.

4. Mix together with shio koji to serve.

 TIPS   You can serve the dish as baby food after cutting it into very small pieces.

Penne with Tomato Sauce

P.54

 INGREDIENTS

20g penne

3 cherry tomatoes, rinsed and chopped

3 black olives, sliced

1/2 pineapple slice, chopped

A dash of chopped herbs

 SEASONINGS

1 tablespoon tomato paste

A pinch of salt

A pinch of sugar

 METHOD

1. Preheat a vacuum flask by filling it with hot water; attach lid and shake gently. Let it sit for a minute before pouring the preheating water out.

2. Put the penne and a pinch of salt into the flask. Fill the flask with boiling water to cover the ingredients. Stir slightly, attach lid and shake gently. Let sit for 8 minutes and drain.

3. Fill the flask again with boiling water to cover the ingredients. Stir slightly, attach lid and shake gently. Let sit for 5 minutes and drain.

4. Add in the tomatoes, black olives and pineapple. Fill the flask with hot water until it's 80% full. Stir slightly, attach lid and shake gently. Let sit for 3 minutes and drain.

5. Stir in seasonings and herbs to serve.

 TIPS   Kids usually love pasta while this tomato pasta is tasty and easy to make with vacuum flasks. You can invite your kids to cook it together with you and enjoy your time in the kitchen!

Spicy and Sour Seafood with Rice Noodles in Soup

P.56

 INGREDIENTS

30g small shrimps, rinsed and deveined

20g squid, rinsed and cut into bite-size

10g clam meat, rinsed

100g round rice noodles

1/2 tomato, hulled and diced

 SEASONINGS

1 tablespoon black vinegar

1/2 tablespoon chili paste

A pinch of salt

A pinch of sugar

 METHOD

1. Preheat a vacuum flask by filling it with hot water; attach lid and shake gently. Let it sit for a minute before pouring the preheating water out.

2. Put the noodles into the flask. Fill the flask with boiling water to cover the ingredients. Stir slightly, attach lid and shake gently. Let sit for 8 minutes and drain.

3. Add in the shrimps, squid and clams. Fill the flask with hot water to cover the ingredients. Stir slightly, attach lid and shake gently. Let sit for 3 minutes and drain.

4. Put the tomato and seasonings into the flask. Fill the flask with hot water until it's 80% full. Stir slightly, attach lid and shake gently. Let sit for a minute and drain.

TIPS Black vinegar helps stimulate your appetite and makes the noodles perfect for summer. Feel free substitute the seafood with vegetables if preferred.

Noodles with Shredded Chicken in Satay Sauce

P.58

 INGREDIENTS

50g chicken, rinsed and shredded

10g carrot, rinsed, peeled and shredded

20g bean sprouts

55g Chinese egg noodles

10g cucumber, rinsed and shredded

A bunch of spring onions, rinsed and shredded

 SEASONINGS

1 tablespoon satay sauce

A drizzle of sesame oil

A pinch of salt

 METHOD

1. Preheat a vacuum flask by filling it with hot water; attach lid and shake gently. Let it sit for a minute before pouring the preheating water out.

2. Put the shredded chicken into the flask. Fill the flask with boiling water to cover the chicken. Stir slightly, attach lid and shake gently. Let sit for 3 minutes and drain.

3. Add in the carrot, bean sprouts and noodles. Fill the flask with hot water until it's 80% full. Stir slightly, attach lid and shake gently. Let sit for 5 minutes and drain.

4. Stir in the cucumber, spring onions and seasonings to serve.

TIPS Udon and Chinese egg noodles are suitable for thermal cooking as they can be cooked through easily. Adding the satay sauce will make the noodles yummier.

Udon with Pork in Curry Sauce

P.60

 INGREDIENTS

60g pork slices, cut into bite-size

30g cabbage, rinsed and chopped

10g yellow bell pepper, rinsed and cut into thin strips

10g red bell pepper, rinsed and cut into thin strips

1/2 udon

 SEASONINGS

1/6 block (20g) Japanese instant curry blocks

A pinch of salt

A pinch of sugar

 METHOD

1. Preheat a vacuum flask by filling it with hot water; attach lid and shake gently. Let it sit for a minute before pouring the preheating water out.

2. Put the pork into the flask. Fill the flask with boiling water to cover the pork. Stir slightly, attach lid and shake gently. Let sit for 3 minutes and drain.

3. Add in the cabbage, bell peppers and udon. Fill the flask with hot water until it's 80% full. Stir slightly, attach lid and shake gently. Let sit for 5 minutes and drain.

4. Stir in seasonings and fill the flask with some water to serve.

 TIPS — Making curry dishes becomes incredibly easy with the easy-to-store Japanese instant curry blocks.

Vietnamese Fish Fillet Cold Vermicelli

P.62

 INGREDIENTS

25g fish, cut into thin slices

50g rice vermicelli noodles

10g bean sprouts

10g carrot, rinsed, peeled and shredded

15g lettuce, rinsed and shredded

A bunch of basil, shredded

A dash of crushed peanuts

 SEASONINGS

1 teaspoon fish sauce

1 teaspoon white vinegar

1/4 teaspoon minced garlic

1/2 teaspoon honey

 METHOD

1. Preheat a vacuum flask by filling it with hot water; attach lid and shake gently. Let it sit for a minute before pouring the preheating water out.

2. Put the fish, noodles and bean sprouts into the flask. Fill the flask with boiling water to cover the ingredients. Stir slightly, attach lid and shake gently. Let sit for 3 minutes and drain. Remove the ingredients from the flask and set aside to let cool.

3. Fill the flask with ice water to cool it down. Attach lid and shake slightly. Let it sit for a minute before pouring the water out.

4. Add in the cooled ingredients, carrot and lettuce. Stir lightly, attach lid and set aside.

5. When ready to serve, stir in the basil, peanuts and seasonings.

 TIPS — The vacuum flask keeps the temperature of its contents stable, so in summer it is one of the best ways to keep your cold dishes cool.

Red Rice with Porcini and Vegetables

P.66

 INGREDIENTS

20g red onion, peeled and cut into thin strips

10g asparagus, rinsed and cut into lengths

10g carrot, peeled and diced

20g red rice

60g white rice

5g porcini mushrooms, soaked to soften, drained and chopped

A pinch of salt

A knob of butter

 METHOD

1. Preheat a vacuum flask by filling it with hot water; attach lid and shake gently. Let it sit for a minute before pouring the preheating water out.

2. Put the onion, asparagus and carrot pieces into the flask. Fill the flask with boiling water to cover the ingredients. Stir slightly, attach lid and shake gently. Let sit for 5 minutes and drain.

3. Add in the rice and fill the flask with hot water until its 80% full. Stir slightly, attach lid and shake gently. Let sit for 30 minutes and drain.

4. Add mushrooms and salt to the flask and fill it with hot water until it's 80% full. Let sit for 90 minutes. When ready to serve, mix together with butter.

TIPS — The rice is fully immersed with the distinctly strong flavor of porcini mushrooms after braising.

Rice with Curry Beef Cubes and Corn

P.68

 INGREDIENTS

30g beef tenderloin, diced

10g sweet corn kernels

10g carrot, peeled and diced

80g white rice

1 teaspoon curry paste

A pinch of salt

 METHOD

1. Preheat a vacuum flask by filling it with hot water; attach lid and shake gently. Let it sit for a minute before pouring the preheating water out.

2. Put the beef into the flask. Fill the flask with boiling water to cover the beef. Stir slightly, attach lid and shake gently. Let sit for 5 minutes and drain.

3. Add in the corns, carrot and rice. Fill the flask with hot water until its 80% full. Stir slightly, attach lid and shake gently. Let sit for 30 minutes and drain.

4. Add curry paste and salt to the flask and fill it with hot water until it's 80% full. Let sit for 90 minutes.

TIPS — If you are cooking for kid, use the Japanese instant curry for children, which taste milder and less salty.

Rice with Five Spice Taro and Sakura Dried Shrimp

 INGREDIENTS

50g taro, peeled and diced

80g white rice

3g sakura dried shrimps

1/2 teaspoon five spice

A pinch of salt

A bunch of shredded dried seaweed

 METHOD

1. Preheat a vacuum flask by filling it with hot water; attach lid and shake gently. Let it sit for a minute before pouring the preheating water out.

2. Put the taro into the flask. Fill the flask with boiling water to cover the taro. Stir slightly, attach lid and shake gently. Let sit for 5 minutes and drain.

3. Add in the rice and fill the flask with hot water until it's 80% full. Stir slightly, attach lid and shake gently. Let sit for 30 minutes and drain.

4. Place the shrimps, five spice and salt into the flask; fill it with hot water until it's 80% full. Stir slightly, attach lid and shake gently. Let sit for 90 minutes and drain. Garnish with seaweed to serve.

 TIPS   The perfect match of five spice and taro together with the sakura shrimps gives the dish a strong "umami" flavor.

Red Beans Rice

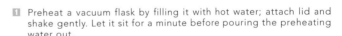

P.72

 INGREDIENTS

10g red beans, soaked to soften and drained

10g Chinese pearl barley, soaked to soften and drained

50g white rice

 METHOD

1. Preheat a vacuum flask by filling it with hot water; attach lid and shake gently. Let it sit for a minute before pouring the preheating water out.

2. Put the red beans and Chinese pearl barley into the flask. Fill the flask with boiling water to cover the ingredients. Stir slightly, attach lid and shake gently. Let sit for 30 minutes and drain.

3. Add in the rice and fill the flask with hot water until it's 80% full. Stir slightly, attach lid and shake gently. Let sit for 30 minutes and drain.

4. Fill the flask with hot water again until it's 80% full. Stir slightly, attach lid and shake gently. Let sit for 90 minutes and it's ready to serve.

 TIPS   The dish is stomach-filling and dampness clearing, while red beans and Chinese pearl barley are great in reducing edema.

Rice with Korean Spicy Squid

P.74

 INGREDIENTS

30g squid, rinsed and cut into bite-size

10g carrot, peeled and shredded

10g bean sprouts

A few pieces of kombu, soaked to soften, drained and cut into thin strips

50g white rice

1 teaspoon Korean hot pepper paste

1/2 teaspoon soy sauce

A drizzle of sesame oil

10g egg strips

 METHOD

1. Preheat a vacuum flask by filling it with hot water; attach lid and shake gently. Let it sit for a minute before pouring the preheating water out.

2. Put the squid pieces into the flask. Fill the flask with boiling water to cover the squid. Stir slightly, attach lid and shake gently. Let sit for 5 minutes and drain.

3. Add in the carrot, bean sprouts, kombu and rice. Fill the flask with hot water until it's 80% full. Stir slightly, attach lid and shake gently. Let sit for 30 minutes and drain.

4. Fill the flask with hot water again until it's 80% full. Stir slightly, attach lid and shake gently. Let sit for 90 minutes. When ready to serve, mix with hot pepper paste, soy sauce, sesame oil and egg strips.

TIPS Feel free to substitute the squid with other seafood.

Diced Chicken and Black Garlic Rice

P.76

 INGREDIENTS

50g chicken, diced

20g onion, peeled and minced

30g red, yellow and green bell peppers (10g each), rinsed, seeded, stem removed and diced

100g white rice

4 cloves black garlic, chopped and minced

A pinch of salt

 METHOD

1. Preheat a vacuum flask by filling it with hot water; attach lid and shake gently. Let it sit for a minute before pouring the preheating water out.

2. Put the diced chicken into the flask. Fill the flask with boiling water to cover the chicken. Stir slightly, attach lid and shake gently. Let sit for 5 minutes and drain.

3. Add in the onion, bell peppers and rice. Fill the flask with hot water until it's 80% full. Stir slightly, attach lid and shake gently. Let sit for 30 minutes and drain.

4. Add black garlic and salt; fill the flask with hot water again until it's 80% full. Stir slightly, attach lid and shake gently. Let sit for 90 minutes.

TIPS Black garlic has a note of dark caramel and adds natural sweetness to the braised rice.

Mixed Sweet Potatoes Congee

P.78

 INGREDIENTS

20g orange sweet potatoes, peeled and diced

20g purple sweet potatoes, peeled and diced

50g white rice

 METHOD

1. Preheat a vacuum flask by filling it with hot water; attach lid and shake gently. Let it sit for a minute before pouring the preheating water out.

2. Put the sweet potatoes into the flask. Fill the flask with boiling water to cover the ingredients. Stir slightly, attach lid and shake gently. Let sit for 5 minutes and drain.

3. Add in the rice and fill the flask with hot water until it's 80% full. Stir slightly, attach lid and shake gently. Let sit for 30 minutes and drain.

4. Fill the flask with hot water again until it's 80% full. Stir slightly, attach lid and shake gently. Let sit for 90 minutes.

 TIPS   Sweet potatoes are high in fiber and nutrition. They can be cooked through easily with thermos cooking, it is a kids-friendly dish too!

Daliang Milk Sheets and Oatmeal Congee

P.80

 INGREDIENTS

1/3 cup white rice

1 tablespoon instant oatmeal

10g ginkgo nuts, shelled, cut into halves and stems removed

2 slices Daliang buffalo cheese slices

 METHOD

1. Preheat a vacuum flask by filling it with hot water; attach lid and shake gently. Let it sit for a minute before pouring the preheating water out.

2. Put the ginkgo nuts into the flask. Fill the flask with boiling water to cover the ingredients. Stir slightly, attach lid and shake gently. Let sit for 5 minutes and drain.

3. Add in the oatmeal and rice. Fill the flask with hot water until it's 80% full. Stir slightly, attach lid and shake gently. Let sit for 30 minutes and drain.

4. Fill the flask with hot water again until it's 80% full. Stir slightly, attach lid and shake gently. Let sit for 90 minutes before stirring in the buffalo cheese to serve.

 TIPS   Daliang buffalo cheese are also known as "Chinese cheese". It adds same savouriness and creaminess to the congee.

Leaf Mustard, Tofu and Fish Soup

P.84

INGREDIENTS

2 ginger slices

Some spring onions, cut into lengths

1 tail of grass carp, rinsed and pat dry

A dash of Shaoxing wine

4 cups water

300g gaichoy (Chinese mustard greens), rinsed and cut into bite-size

30g bean sprouts

360g tofu, diced

A pinch of salt

METHOD

1 Add ginger slices and spring onions into the pot of a vacuum cooker and stir-fry until fragrant on stove. Put the grass carp in and pan-fry until the fish is lightly browned.

2 Deglaze the pot with Shaoxing wine and boiling water.

3 Place gaichoy, bean sprouts and tofu in the pot. When boiled, attach lid and cook over middle heat for 5 minutes. Remove the pot from heat and put it into the vacuum flask. Let sit for 30 minutes. Season with salt to taste.

TIPS To make the soup milky, make sure you add boiling water when the fish in the pot is lightly browned. The soup helps reduce excessive heat from your body. For busy HongKongers who are always too busy to cook, this simple recipe really helps make cooking much easier!

Sea Coconut and Lily Bulb Soup

P.86

INGREDIENTS

100g lean pork

100g fresh sea coconut, peeled, rinsed and cut into halves

10g dried lily bulbs

1 carrot, peeled, rinsed and cut into pieces

1 whole corn, peeled, rinsed and cut into chunks

1 apple, cut into bite-size and seeded

2 dried mutcho dates

6 cups water

METHOD

1 Blanch the pork in boiling water and drain.

2 Put all ingredients into the pot of a vacuum cooker and cook until boiled. Attach lid and continue to cook over middle heat for 5 minutes. Remove the pot from heat and put it into the vacuum flask. Seal and let sit for 2 hours before serving.

TIPS Sea Coconut meat is white and tender with a mild sweetness. It has the effect of relive coughing, clear up lungs and reduce heat from bodies in the view of Chinese medical perception. It is a great soup for kids, especially during fall.

Beetroot and Cashew Soup

 INGREDIENTS

200g beets, peeled and cut into bite-size

100g carrot, peeled and cut into chunks

130g chestnuts, shells and inner skin removed

1 whole corn, peeled, rinsed and cut into chunks

1 pear, peeled, cut into pieces and seeded

60g cashew nuts

2 dried shitake mushrooms

20g snow fungus, soaked to soften, drained and hard ends removed

2 dried mutcho dates

6 cups water

 METHOD

P.88

1 Put all ingredients into the pot of a vacuum cooker and cook until boiled. Attach lid and continue to cook over middle heat for 5 minutes. Remove the pot from heat and put it into the vacuum flask. Seal and let sit for 2 hours before serving.

 TIPS The tip for making a vegetarian soup rich and favorable is to add a lot of cashews and dried mushrooms to the soup; while the nuts also make the soup more stomach-filling.

Cordyceps Flower Soup with Chinese Yam , Lotus Seed and Sea Whelk

 INGREDIENTS

20g cordyceps flowers, soaked to soften and drain

40g dried Chinese yam, rinsed

30g dried lotus seed, rinsed and germs removed

30g gordon euryale seed, rinsed

400g frozen sea whelks, rinsed

250g lean pork

10 cups water

A pinch of salt

 METHOD

P.90

1 Blanch sea whelks and pork in boiling water. Drain and set aside.

2 Put all ingredients (except for the salt) into the pot of a vacuum cooker and cook until boiled. Attach lid and continue to cook over middle heat for 5 minutes. Remove the pot from heat and put it into the vacuum flask. Seal and let sit for 2 hours. Season with salt to serve.

Mangosteen, Corn and Pork Bone Soup

 INGREDIENTS

1 luohanguo (monk fruit), slightly crushed

1 whole corn, peeled, rinsed and cut into chunks

1 carrot, peeled and cut into pieces

300g pork cheek bones

8 water chestnuts, peeled

2 dried mutcho dates

6 cups water

A pinch of salt

 METHOD

P.92

1 Blanch the pork cheek bones. Rinse and drain.

2 Put all ingredients (except for the salt) into the pot of a vacuum cooker and cook until boiled. Attach lid and continue to cook over middle heat for 5 minutes. Remove the pot from heat and put it into the vacuum flask. Seal and let sit for 2 hours. Season with salt to serve.

 TIPS Pork cheek bone is a meaty cut which has a similar texture as beef shank's. I love using it in my soups as it is soft and tender after cooking, and it also brings natural sweetness to the soup.

Chiu Chow Beef Brine

P.96

 INGREDIENTS

1kg frozen beef shank, excess fat removed with knife

 SPICES

5 dried bay leaves

10g anise

2g Chinese cinnamon bark

1 teaspoon white peppercorns

Some cloves

2 sand ginger slices

4 black cardamom

5g licorice root

 SAUCE

3 garlic cloves, peeled

3 ginger slices

2 shallots, peeled and cut into chunks

A dash Shaoxing wine

30g rock sugar

1 tablespoon oyster sauce

3 tablespoon soy sauce

1 tablespoon dark soy sauce

1 teaspoon salt

6 cups water

 METHOD

1. Put the spices into a filter bag and tie it shut. Place the bag, beef shank and the ingredients for the sauce into the pot of a thermal cooker. Cook until boiled, attach lid and continue to cook over middle heat for 20 minutes.

2. Remove from heat and put the pot into the vacuum flask. Seal and let sit for 90 minutes. When done, remove the shank from the pot and cut it into slices to serve.

 TIPS

You would love the recipe if you enjoy the chewy texture of beef shanks. But if you prefer a more tender texture and stronger flavor, put the beef slices back to the pot and cook on stove again. When boiled, attach lid and continue to cook over middle heat for 5 minutes. Then put the pot into the vacuum flask; seal and let sit for another 30 minutes.

Stewed Chicken Wings in Goji Berry and Plum Sauce

P.98

 INGREDIENTS

10 chicken wings

2 garlic cloves, crushed and peeled

1/2 red onion, peeled and cut into chunks

1/2 apple, cut into bite-size and seeded

5 dried plums

Some rock sugar

1 tablespoon dark soy sauce

2 tablespoons soy sauce

300ml water

1 tablespoon dried goji berries, soaked to soften and drain

 METHOD

1. Heat oil in the pot of a thermal cooker on stove. Place chicken wings in the pot, cook until golden brown on both sides. Remove chicken wings from the pot and set aside.

2. Add the garlic and red onion to the pot and stir fry until fragrant. Put the wings back to the pot; cook together with the apple, plums, rock sugar, soy sauce and water. When boiled, attach lid and continue to cook over middle heat for 5 minutes.

3. Stir in goji berries and remove from heat. Put the pot into the vacuum flask; seal and let sit for 30 minutes.

 TIPS

Braised chicken wings taste less oily than sautéed wings. And in this recipe, the dried plums add a pleasantly sour touch to the dish and make it more appetizing. It sure serves a great company with rice!

Stewed Pork Cartilage with Lotus in Korean Sauce

P.100

 INGREDIENTS

300g pork cartilage

2 tablespoon Korean meat marinade (bulgogi sauce)

1/2 onion, peeled and cut into chunks

120g carrot, peeled and cut into pieces

350g lotus root, peeled and cut into bite-size

350ml water

 SEASONINGS

1 teaspoon minced garlic

1 tablespoon Korean red pepper paste

1 tablespoon white vinegar

1 teaspoon sugar

 METHOD

1. Mix together pork and marinade in a bowl.

2. Heat oil in the pot of a thermal cooker on stove. Place the pork in the pot and cook until golden brown. Remove the pork from the pot and set aside.

3. Add the onion and carrot to the pot and cook until fragrant. Put the lotus root in the pot and stir slightly. Then add the pork back and stir in seasonings. Pour in water and cook over middle heat; when boiled, attach lid and continue to cook for 10 minutes. Remove the pot from heat and put it into the vacuum flask; seal and let sit for 90 minutes.

 TIPS The soft but chewy braised pork cartilage combined with the crunchy lotus root creates a uniquely complex texture for the dish, which makes it stand out from other braised pork cartilage dishes.

Five Spice Beef Brisket and Bean Curd Sheets in Hot Pot

P.102

 INGREDIENTS

1kg beef brisket, cut into bite-size

3 ginger slices

Some spring onions cut into lengths

500ml water

600g white radish, cut into bite-size

1 long red pepper

200g dried bean curds

8 tofu puffs

 SEASONINGS

1/2 teaspoon five spice powder

A pinch of salt

A pinch of black pepper powder

3 tablespoons chu hou paste

2 tablespoons oyster sauce

Some rock sugar

 METHOD

1. Mix together the beef brisket, five spice powder, salt and black pepper in a bowl.

2. Heat oil in the pot of a thermal cooker on stove. Add the ginger and spring onions to the pot and cook until fragrant. Place the beef in the pot and cook until golden brown. Stir in the chu hou paste.

3. Add in water, oyster sauce, sugar, radish and red pepper. Cook until boiled. Attach lid and continue to cook over medium heat for 20 minutes.

4. Stir in bean curds and tofu puffs. Remove the pot from heat and put it into the vacuum flask; seal and let sit for 90 minutes.

 TIPS The hearty and delicious dish is ideal for serving in winter, but is time and energy consuming to cook on stove. Therefore it is better to cook it with a vacuum cooker, which saves both time and energy, and reduces the carbon footprint at the same time.

Mushroom, Yam and Chestnut

 INGREDIENTS

100g chestnuts

400g chicken, chopped chunky, rinsed and drained

2 shallots, peeled and cut into pieces

2 garlic cloves, crushed and peeled

120g carrot, peeled and cut into chunks

30g dried shiitake mushrooms, soaked to soften, drained, hard ends removed and sliced

150g fresh Chinese yam, peeled and cut into bite-size

100g broccoli, cut into florets and rinsed

300ml water

 MARINADE

1 tablespoon soybean paste

1 teaspoon rice wine

A dash of sesame oil

 METHOD

1. Boil chestnut in a pot with enough water to cover them over middle heat. When boiled, remove from heat and pour in cold water to cool down. Peel the shell and inner skin of the chestnuts with a knife. Set the chestnuts aside.

2. Mix the chicken with marinade in a bowl, set aside.

3. Heat oil in the pot of a thermal cooker on stove. Add the shallots and garlic cloves to the pot and cook until fragrant. Put the chicken in and cook until golden brown. Then add the carrot and stir slightly.

4. Add in the chestnuts, mushrooms, yam, broccoli, and pour in water. When boiled, attach lid and continue to cook over middle heat for 5 minutes. Remove the pot from heat and put it into the vacuum flask; seal and let sit for 45 minutes.

 TIPS Cooking the dish on stove for 5 minutes then braising it in the thermal cooker for 45 minutes ensures that all the flavors and nutrition of the food will be entrapped in the cooker.

Chicken with Dried Porcini in Soy Sauce

P.106

 INGREDIENTS

1 whole chicken, rinsed

2 ginger slices

3 shallots, peeled and cut into pieces

60g carrot, peeled and cut into pieces

20g porcini mushrooms, soaked to soften and drained

50g dried shiitake mushrooms, soaked to soften, drained, hard ends removed and sliced

50g beech mushrooms, base of the cluster trimmed

 MARINADE

1 teaspoon mei kuei lu chiew (rose cooking wine)

1 tablespoon dark soy sauce

2 tablespoons soy sauce

 SAUCE

60ml soy sauce

2 tablespoons dark soy sauce   2 bay leaves

50g rock sugar   14 cups water

 METHOD

1. Mix together all ingredients of the marinade and spread it evenly on the chicken. Let sit for 15 minutes.

2. Heat oil in the pot of a thermal cooker on stove. Add the shallots, ginger slices and carrot to the pot and cook until fragrant. Put the mushrooms in and stir slightly.

3. Place the chicken and the marinade into the pot and cook until boiled. Attach lid and continue to cook over middle heat for 10 minutes. Remove the pot from heat and put it into the vacuum flask; seal and let sit for 60 minutes.

 TIPS The mushrooms lend a refreshing aroma to the dish, so the dish best suits the taste buds of modern people that pay attention to healthy eating.

Osmanthus, Apple, White Pear and Honey Locust Fruit Sweet Soup

 INGREDIENTS

1 tablespoon dried osmanthus

60g honey locust fruit, soaked overnight to soften and drained

2 apples, rinsed, cut into pieces and seeded

2 pears, rinsed, cut into pieces and seeded

12 cups water

60g rock sugar

 METHOD

1 Put everything (except rock sugar) into the pot of a thermal cooker. Attach lid and cook on stove until boiled. Then continue to cook over middle heat for 5 minutes.

2 Add in rock sugar and stir until melted. Attach lid and cook over middle heat until boiled. Remove from heat and put the pot into the vacuum flask. Seal and let sit for 30 minutes.

TIPS Honey locust fruit have great nutritional values and are regarded as "vegetarian bird's nest" due to their beauty benefits.

Dried Beancurd, Chinese Pearl Barley and Peach Gum Sweet Soup

P.112

 INGREDIENTS

15g bean curd sheets

40g peach gum, soaked to soften, rinsed and drained

90g rock sugar

30g Chinese pearl barley

6 cups water

 METHOD

1 Put everything (except rock sugar) into the pot of a thermal cooker. Attach lid and cook on stove until boiled. Then continue to cook over middle heat for 10 minutes.

2 Add in rock sugar and stir until melted. Attach lid and cook over middle heat until boiled. Remove from heat and put the pot into the vacuum flask. Seal and let sit for 90 minutes.

TIPS Peach gum is the resin secreted from the bark of peach tree. It is high in plant collagen with a mild nature. Its jelly-like texture gives some bite to the nourishing and heat-clearing sweet soup.

Egg, Sweet Purple Potato and Rice Cake Sweet Soup

P.114

 INGREDIENTS

320g purple sweet potatoes, peeled and cut into bite-size

150g rice cake

140g rock sugar

8 cups water

1 egg

 METHOD

1 Boil water in the pot of a thermal cooker on stove. Put sweet potatoes, rice cake and sugar into the pot, stir well. Attach lid and cook until boiled. Then continue to cook over middle heat for 3 minutes.

2 Remove the pot from heat and put it into the vacuum flask. Seal and let sit for 15 minutes. Stir in the beaten egg for egg drop soup.

TIPS Cooking rice cake in a thermal cooker helps retain its chewy texture; while the purple sweet potatoes add great flavor and color to the dish.

Fig, White Pear and Lily Bulb Congee

P.116

 INGREDIENTS

2 dried figs, cut into halves

1 pear, peeled, cut into bite-size and seeded

3/4 cup white rice (of a rice measuring cup)

6 cups water

2 fresh lily bulbs, rinsed and torn into small pieces

1 tablespoon goji berries, soaked until soften and drained

30g rock sugar

 METHOD

1 Boil the figs, pear, rice and water in the pot of a thermal cooker on stove with a lid on. Continue to cook over middle heat for 5 minutes.

2 Stir in lily bulbs, goji berries and sugar. Attach lid and cook until boiled. Remove the pot from heat and put it into the vacuum flask. Let sit for 30 minutes.

 TIPS   The congee is mild and nourishing, which suits both adults and children.

Tomato, Shrimp and Five Grains Congee

P.118

 INGREDIENTS

1 teaspoon minced garlic

1/2 onion, peeled and diced

6 prawns

2 tomatoes, hulled and diced

2 tablespoon tomato paste

1 teaspoon salt

1 teaspoon sugar

1 1/2 cup white rice (of a rice measuring cup)

2 tablespoons oatmeal

1 tablespoon mixed quinoa (black, red and white quinoa mix)

6 cups water

 METHOD

1 Heat oil in the pot of a thermal cooker on stove. Place the garlic and onion in the pot, cook until fragrant. Sauté prawns until cooked.

2 Stir in tomatoes and tomato paste, then season with salt and sugar. Mix together with rice, oatmeal and quinoa. Pour in water and attach lid. Cook over middle heat on stove for 5 minutes. Remove the pot from heat and put it into the vacuum flask. Let sit for 30 minutes.

 TIPS   Sauteing the ingredients before boiling creates maximum flavor of the dish and make it more tasteful.

10分鐘OK燜燒料理

作者	梁雅琳
總編輯	Ivan Cheung
責任編輯	Sophie Chan
助理編輯	Tessa Tung
文稿校對	Joyce Leung, Grace Chan
封面設計	SO@SmilePro
內文設計	Eva
出版	研出版 In Publications Limited
市務推廣	Samantha Leung
查詢	info@in-pubs.com
傳真	3568 6020
地址	九龍太子白楊街 23 號 3 樓
香港發行	春華發行代理有限公司
地址	香港九龍觀塘海濱道 171 號申新證券大廈 8 樓
電話	2775 0388
傳真	2690 3898
電郵	admin@springsino.com.hk
台灣發行	永盈出版行銷有限公司
地址	新北市新店區中正路505號2樓
電話	886-2-2218-0701
傳真	886-2-2218-0704
出版日期	2017 年 03 月 30 日
ISBN	978-988-77349-3-2
售價	港幣 $98 / 新台幣 $430